IBM SPSS for Introductory Statistics

IBM SPSS for Introductory Statistics is designed to help students learn how to analyze and interpret research. In easy-to-understand language, the authors show readers how to choose the appropriate statistic based on the design, and to interpret outputs appropriately. There is such a wide variety of options and statistics in SPSS that knowing which ones to use and how to interpret the outputs can be difficult. This book assists students with these challenges.

Comprehensive and user-friendly, this book prepares readers for each step in the research process: design, entering and checking data, testing assumptions, assessing reliability and validity, computing descriptive and inferential parametric and nonparametric statistics, and writing about results. Dialog windows and SPSS syntax, along with the output, are provided. Several realistic datasets are used to solve the chapter problems and are available as an online resource.

This edition includes the following:

- Updated chapters and screenshots
- Additional SPSS work problems
- Callout boxes for each chapter, indicating crucial elements of APA style and referencing outputs

IBM SPSS for Introductory Statistics is an invaluable supplemental (or lab text) book for students. In addition, this book and its companion, IBM SPSS for Intermediate Statistics, are useful as guides/reminders to faculty and professionals regarding the specific steps to take to use SPSS and/or how to use and interpret parts of SPSS with which they are unfamiliar.

Karen C. Barrett is Professor Emerita of Human Development and Family Studies at Colorado State University, where she taught research methods and statistics classes as well as classes in her research area. She is also Professor of Community & Behavioral Health at Colorado School of Public Health. She received her PhD in developmental psychology from the University of Denver. Her research takes a functional approach to studying emotional and motivational processes and their influence on development; family and cultural influences on emotion regulation; and the development of emotion regulation and social emotions.

Nancy L. Leech is Professor of Research and Evaluation Methods at the University of Colorado, Denver. She teaches graduate level courses in research, statistics, and measurement. She received her PhD in education with an emphasis on research and statistics from Colorado State University in 2002. Her area of research is promoting new developments and better understandings in applied, quantitative, qualitative, and mixed methods research.

Gene W. Gloeckner is Professor Emeritus. He has served as IRB Chair, School of Education Director, and Semester at Sea Dean. He received the Jack E Cermak University Advising Award in 2024. He earned his PhD and BS from The Ohio State University and MS from Colorado State University. Much of his writing and teaching has focused on issues in quantitative and mixed research methods. He has served as the academic advisor for over 75 doctoral graduates.

George A. Morgan is Emeritus Professor of Education and Human Development at Colorado State University. He received his PhD in child development and psychology from Cornell University. In addition to writing textbooks, he has advised many PhD students in education and related fields. He has conducted a program of research on children's motivation to master challenging tasks.

IBM SPSS for Introductory Statistics

Use and Interpretation

Seventh Edition

Karen C. Barrett, Nancy L. Leech, Gene W. Gloeckner and George A. Morgan

Routledge
Taylor & Francis Group

NEW YORK AND LONDON

Cover image: jadamprostore via Getty Images

Seventh edition published 2026
by Routledge
605 Third Avenue, New York, NY 10158

and by Routledge
4 Park Square, Milton Park, Abingdon, Oxon, OX14 4RN

Routledge is an imprint of the Taylor & Francis Group, an informa business

© 2026 Karen C. Barrett, Nancy L. Leech, Gene W. Gloeckner and George A. Morgan

First edition published by Erlbaum Associates 2001
Sixth edition published by Routledge 2020

Library of Congress Cataloging-in-Publication Data
A catalog record for this book has been requested

ISBN: 978-1-032-41031-9 (hbk)
ISBN: 978-1-032-41030-2 (pbk)
ISBN: 978-1-003-35590-8 (ebk)

DOI: 10.4324/9781003355908

Typeset in Times New Roman
by Apex CoVantage, LLC

Screen image reprints courtesy of IBM Corporation ©2025. IBM, the IBM logo, ibm.com, and SPSS are trademarks or registered trademarks of International Business Machines Corporation, registered in many jurisdictions worldwide. Used with permission. This book is not approved or sponsored by SPSS Statistics.

Access the Support Material: www.routledge.com/9781032410302

Contents

Acknowledgments

This IBM SPSS book is consistent with and could be used as a supplement for Gliner et al. (2017), *Research Methods in Applied Settings: An Integrated Approach to Design and Analysis* (2nd ed.), which provides extended discussions of how to conduct a quantitative research project as well as understand the key concepts. Or this SPSS book could be a supplement for Morgan et al. (2006), *Understanding and Evaluating Research in Applied and Clinical Settings*, which is a shorter book emphasizing reading and evaluating research articles and statistics. Information about both books can be found at www.psypress.com.

Because this book draws heavily on these two research methods texts and on earlier editions of this book, we need to acknowledge the important contribution of three current and former colleagues. We thank Jeff Gliner for allowing us to use material in Chapters 1, 2, and 3. Bob Harmon facilitated much of our effort to make statistics and research methods understandable to students, clinicians, and other professionals. We hope this book will serve as a memorial to him and the work he supported. Orlando Griego was a co-author of the first edition of this SPSS book; it still shows the imprint of his student-friendly writing style.

We would like to acknowledge the assistance of the many students who have used earlier versions of this book and provided helpful suggestions for improvement. We could not have completed the current version of the book or made it look so good without the assistance of Marisha Lamont-Manfre or Xia Xue and prior versions without Don Quick, Sonia Nelson, Linda White, Catherine Lamana, Alana Stewart, Sophie Nelson, Jikyeong Kang, Bill Sears, LaVon Blaesi, Mei-Huei Tsay, Sheridan Green, Lisa Vogel, Andrea Weinberg, Pam Cress, Joan Clay, Laura Jensen, James Lyall, Joan Anderson, Yasmine Andrews, Jessica Bochert, and Jessica Gerton. Bob Fetch and Ray Yang provided helpful feedback on the readability and user friendliness of earlier versions of the text. Finally, the patience of our spouses (Terry, Grant, Susan, and Hildy) and families enabled us to complete the task without too much family strain.

The screenshots of the many SPSS windows are reprinted by courtesy of International Business Machines Corporation, © SPSS, Inc., an IBM Company. SPSS was acquired by IBM in October 2009.

Preface

This book is designed to help students learn how to analyze and interpret research. It is intended to be a supplemental text in an introductory (undergraduate or graduate) statistics or research methods course in the behavioral or social sciences or education, and it can be used in conjunction with any mainstream text. We have found that this book makes IBM SPSS Statistics for Windows easy to use so that it is not necessary to have a formal, instructional computer lab; however, it is also suitable as a text to use in such a lab if the course has one. Access to the IBM SPSS program and some familiarity with Windows is all that is required. Although SPSS is quite easy to use, there is such a wide variety of options and statistics that knowing which ones to use and how to interpret the printouts can be difficult. This book is intended to help with these challenges. In addition to serving as a supplemental or lab text, this book and its companion, *IBM SPSS for Intermediate Statistics* (Barrett, Leech, & Morgan, 6th ed., 2025), are useful as reminders to faculty and professionals of the specific steps to take to use SPSS and/or guides to using and interpreting parts of SPSS with which they might be unfamiliar.

The Computer Program

We used IBM SPSS 29 in this book. Although some changes in SPSS have occurred across different versions, most of the instructions in this book can be applied to prior versions. We expect future Windows versions of this program to be similar so students should be able to use this book with earlier and later versions of the program, which we call SPSS in the text. Our students have used this book, or earlier editions of it, with all of the earlier versions of SPSS; both the procedures and outputs are quite similar. We point out some of the changes at various points in the text.

In addition to various SPSS modules that may be available at your university, there are two versions that are available for students that you can rent for 6 or 12 months online. *Statistics GradPack* enables you to do all the statistics in this book plus most of those in our *IBM SPSS for Intermediate Statistics* book (Barrett et al., 2025) and many others (however, there are a few statistics that are not included, so you may want to check before deciding to purchase).

Goals of This Book

Helping you learn how to choose the appropriate statistics, interpret the outputs, and develop skills in writing about the meaning of the results are the main goals of this book. Thus, we have included material on the following:

1. How the appropriate choice of a statistic is influenced by the design of the research.
2. How to use SPSS to help the researcher answer research questions.

3. How to interpret SPSS outputs.
4. How to write about the outputs in the Results section of a paper.

This information will help you develop skills that cover the whole range of the steps in the research process: design, data collection, data entry, data analysis, interpretation of outputs, and writing results. The modified high school and beyond dataset (HSB) used in this book is similar to one you might have for a thesis, dissertation, or research project. Therefore, we think it can serve as a model for your analysis.

The website www.routledge.com/9781032410302 contains the HSB data files under the Student Resources tab used throughout this book (hsbdata and hsbdataNEW). Two other datasets (called CollegeStudentData.sav and ChapterFourData.sav) are used for the extra statistics problems at the end of most chapters, and DataFastTrack.sav and DataRegularTrack.sav are used in Appendix A for the merging of two data files. Appendix A shows how to download these files from the website to your computer.

This book demonstrates how to produce a variety of statistics that are usually included in basic statistics courses, plus others (e.g., reliability measures) that are useful for doing research. We try to describe the use and interpretation of these statistics as much as possible in nontechnical, jargon-free language. In part, to make the text more readable, we have chosen not to cite many references in the text; however, we have provided a short bibliography, "For Further Reading," of some of the books and articles that our students have found useful. We assume that most students will use this book in conjunction with a class that has a textbook; it will help you to read more about each statistic before doing the assignments.

Overview of the Chapters

Our approach in this book is to present how to use and interpret the SPSS statistics program in the context of proceeding as if the HSB data were the actual data from your research project. These chapters are organized in very much the same way you might proceed if this were your project. The goal is to use this computer program as a tool to help you answer these research questions. (Appendix B provides some guidelines for phrasing or formatting research questions.) Chapter 2 provides an introduction to data coding, entry, basic transformations to turn the raw data into variables, how to check data for errors, and how to do descriptive statistics. We developed Chapter 2 because many of you may have little experience with making "messy," realistic data ready to analyze and finding out its measurement properties. In this revision, we mainly focused on how to use SPSS to analyze online data, but we do provide some guidance for those who must enter data by hand. Chapter 3 provides a brief overview of research designs (e.g., between groups and within subjects). This chapter also provides flowcharts and tables useful for selecting an appropriate statistic. Also included is an overview of how to interpret and write about the results of an inferential statistic. This includes not only information about testing for statistical significance but also a discussion of effect size measures and guidelines for interpreting them.

Chapter 4 provides examples of how to check your data for evidence of reliability and validity using several statistics provided by SPSS; for example, Cohen's kappa and Cronbach's alpha. The chapter also provides an introduction to exploratory factor analysis used to reduce a large number of variables to a more manageable number.

Chapters 5 through 7 are designed to answer the several research questions posed in Chapter 1 as well as a number of additional questions. Solving the problems in these chapters should give you a good idea of the basic statistics that can be computed with this computer program. Hopefully, seeing how the research questions and design lead naturally to the choice of statistics will become apparent after using this book. In addition, it is our hope that interpreting what you get back from the computer will become clearer after doing these assignments, studying the outputs, answering the interpretation questions, and doing the extra statistics problems.

Our Approach to Research Questions, Measurement, and Selection of Statistics

In Chapters 1, 2, and 3, our approach is somewhat nontraditional because we have found that students have a great deal of difficulty with some aspects of research and statistics but not others. Most can "crunch" the numbers quite easily and accurately with a computer. However, many have trouble knowing what statistics to use and how to interpret the results. They do not seem to have a "big picture" or see how research design and measurement influence data analysis. Part of the problem is inconsistent terminology. We are reminded of Bruce Thompson's frequently repeated, intentionally facetious remark at his many national workshops: "We use these different terms to confuse the graduate students." For these reasons, we have tried to present a semantically consistent and coherent picture of how research design leads to three basic kinds of research questions (difference, associational, and descriptive) that, in turn, lead to three kinds or groups of statistics with the same names. We realize that these and other attempts to develop and utilize a consistent framework are both nontraditional and somewhat of an oversimplification. However, we think the framework and consistency pay off in terms of student understanding and ability to actually use statistics to help answer their research questions. Instructors who are not persuaded that this framework is useful can skip or modify Chapters 1, 2, and 3 and still have a book that helps their students use and interpret SPSS.

Major Changes in This Edition

The major changes in this edition are based on extensive feedback from students in our classes. Based on this feedback, we focused in Chapter 2 on how to download data collected online and how to upload it to SPSS and transform "string data" (words, rather than numbers) often obtained using free online platforms to numerical data that can be analyzed using SPSS. We moved the information on how to do basic transformations of data and to do descriptive statistics on data to Chapter 2 as well, so that the new Chapter 2 enables the reader to take all of the most commonly needed steps to get data ready to analyze in SPSS to answer inferential research questions. We also have added the relevant parts of the inferential statistic selection chart and effect size chart from Chapter 3 to each chapter that involves inferential statistics. We have included more information about why we make the choices we do in each computer problem in the book. In addition, we now have only one chapter on basic (one dependent variable) difference question inferential statistics, so that students and researchers can find whichever analysis they need in the same chapter, and we have updated the information on which programs calculate effect sizes for difference statistics. We also updated the windows and text to IBM SPSS 29, and we have attempted to correct any typos in the 6th edition and clarify some passages. Although this edition of our *IBM*

SPSS for Introductory Statistics was written using version 29, the program is sufficiently similar to prior versions of this software that we feel you should be able to use this book with earlier and later versions as well.

Instructional Features

Several user-friendly features of this book include the following:

1. Both words and the key **windows** that you see when performing the statistical analyses. This has been helpful to "visual learners."
2. The **outputs** for the analyses that we have done so you can see what you will get (we have done some editing, as shown in Appendix A, to make the outputs fit better on the pages).
3. **Callout boxes** on the outputs that point out parts of the output to focus on and indicate what they mean.
4. For each output, a boxed **interpretation section** that will help you understand the output.
5. Chapter 3 provides specially developed flowcharts and tables to help you **select an appropriate inferential statistic** and **interpret statistical significance and effect sizes**. This chapter also provides an extended example of how to identify and write a research problem, research questions, and a results paragraph.
6. For the statistics in Chapters 5–7, an example of **how to write about the output** and make a table or figure for a thesis, dissertation, or research paper using the 7th edition (2020) of the *Publication Manual of the American Psychological Association* is provided.
7. **Interpretation questions** for each chapter that stimulate you to think about the information in the chapter.
8. Several **Extra Problems** at the end of each chapter for you to run with the SPSS program.
9. Appendix A provides information about how to get started with SPSS and how to use several commands not discussed in the chapters.
10. Appendix B provides examples of how to **write research problems** and **research questions or hypotheses**.
11. **Answers** to the odd numbered **interpretation questions are provided in** Appendix C.
12. **Datasets on the book webpage** www.routledge.com/cw/morgan are available and are listed in Appendix A. These six realistic datasets provide you with data to be used to solve the chapter and Appendix A problems and the end of chapter Extra SPSS Problems.
13. A **Resource Website** is available to students and instructors. To access the site please visit www.routledge.com/9781032410302. Some of the material is password protected and available only to instructors to aid them in teaching the course. Instructors will find the following items available for each chapter: PowerPoint slides, Additional Activities/Suggestions for Instructors, and the answers to the even-numbered Interpretation Questions found in the book (the odd answers are in the book itself). Both students and instructors can access the following material that is provided for each chapter: Chapter Study Guides, Extra SPSS Problems, and Chapter Outlines. Students and instructors, as well as researchers who purchase copies for their personal use, can also access the data files by visiting www.routledge.com/9781032410302.

Major Statistical Features of This Edition

Based on our experiences using the book with students, feedback from reviewers and other users, and the revisions in policy and best practice specified by the APA Task Force on Statistical Inference (1999) and the 6th edition of the *APA Publication Manual* (2010), we have included discussions of the following:

1. **Effect size.** We discuss effect size in each interpretation section to be consistent with the requirements of the revised APA manual. Because this program doesn't provide effect sizes for all the demonstrated statistics, we often have to show how to estimate or compute them by hand.
2. **Writing about outputs.** We include examples of how to write about and make APA-type tables from the information in the outputs. We have found the step from interpretation to writing quite difficult for students so we put emphasis on writing research results.
3. **Data entry and checking.** Chapter 2 describes how the researcher or student looks for errors or inconsistencies in their data and how to make decisions about how to code data. We hope this quite realistic task will help students be more sensitive to issues of data checking *before* doing analyses.
4. **Descriptive statistics and testing assumptions.** In Chapter 2 we also describe exploratory data analysis (EDA), how to test assumptions, and data file management.
5. **Assumptions.** When each inferential statistic is introduced in Chapters 5–7, we have a brief section about its assumptions and when it is appropriate to select that statistic for the problem or question at hand.
6. All the **basic descriptive and inferential statistics such as chi-square, correlation, *t* tests, and one-way ANOVA** covered in basic statistics books. Our companion book, Barrett et al. (2025), *IBM SPSS for Intermediate Statistics: Use and Interpretation* (6th ed.), also published by Routledge/Taylor & Francis, is on the "For Further Reading" list at the end of this book. We think that you will find it useful if you need more complete examples and interpretations of complex statistics including but not limited to **Cronbach's alpha**, **factor analysis**, **multiple regression**, and **factorial ANOVA** that are introduced briefly in this book, as well as many that are beyond the scope of this book.
7. **Reliability and validity assessment.** We present some ways of assessing reliability and validity in Chapter 4. More emphasis on reliability, validity, and testing assumptions is consistent with our strategy of presenting computer analyses that students would use in an *actual* research project.
8. **Nonparametric statistics.** We include the nonparametric tests that are similar to the *t* tests (Mann-Whitney and Wilcoxon) and single factor ANOVA (Kruskal-Wallis) in appropriate chapters, as well as several nonparametric measures of association. This is consistent with the emphasis on checking assumptions because it provides alternative procedures for the student when key assumptions are markedly violated.
9. **SPSS syntax.** We show the syntax along with the outputs because a number of professors and skilled students like seeing and prefer using syntax to produce outputs. How to include SPSS syntax in the output and to save and reuse it is presented in Appendix A. Use of syntax to write

commands not otherwise available in SPSS is presented briefly in our companion volume, Barrett et al. (2025).

Bullets, Arrows, Bold, and Italics

To help you do the problems, we have developed some conventions. We use bullets to indicate actions in SPSS windows that you will take. For example:

- Highlight *academic track* and *math achievement.*
- Click on the arrow to move the variables into the right-hand box.
- Click on **Options** to get Figure 2.16.
- Check **Mean**, **Std Deviation**, **Minimum**, and **Maximum**.
- Click on **Continue**.

Note that the words in italics are variable names and words in bold are words that you will see in the windows and utilize to produce the desired output. In the text they are spelled and capitalized as you see them in the windows. Bold is also used to identify key terms when they are introduced, defined, or important to understanding.

To access a window from what SPSS calls the **Data View** (see Chapter 2), the words you will see in the pull down menus are given in bold with arrows between them. For example:

- Select **Analyze → Descriptive Statistics → Frequencies**.

(This means pull down the Analyze menu, then slide your cursor down to Descriptive Statistics and over to Frequencies, and click.)

Occasionally, we have used underlines to emphasize critical points or commands.

We have tried hard to make this book accurate and clear so that it could be used by students and professionals to learn to compute and interpret statistics without the benefit of a class. However, we find that there are always some errors and places that are not totally clear. Thus, we would like for you to help us identify any grammatical or statistical errors and to point out places that need to be clarified. Please send suggestions to *Karen.Barrett@colostate.edu.*

1 Variables, Research Problems, and Questions

Research Problems

The research process begins with an issue or problem of interest to the researcher. This **research problem** is a statement <u>about the relationships between two or more variables</u>; however, almost all research studies have *more* than two variables.[1] Typically, the research problem is subdivided into a set of more specific questions about the relations among two or more variables. Appendix B provides templates to help you phrase your research problem and different types of research questions. It also provides examples from the expanded high school and beyond (HSB) dataset that is described in this chapter and used throughout the book.

The process of moving from a sense of curiosity, or a feeling that there is an unresolved problem to a clearly defined, researchable problem, can be complex and long. That part of the research process is beyond the scope of this book, but it is discussed in most books about research methods and books about completing a dissertation or thesis.

Variables

Key elements in a research problem are the variables. A **variable** is defined as a characteristic of the participants or situation in a given study that has different values. A <u>variable must vary or have different scores/values in the study</u>. For example, *age* is a variable that can have a large number of values. *Type of treatment/intervention* (or *type of curriculum*) is a variable if there is more than one treatment or a treatment and a control group. The *number of days* to learn something or to recover from an ailment are common measures of the effect of a treatment and, thus, are also potential variables. Similarly, *amount of mathematics knowledge* can be a variable because it can vary from none to a lot.

However, even if a characteristic has the potential to be a variable, if it has only one value in a particular study, it is not a variable; it is a constant. Thus, ethnic group is not a variable if all participants in the study are Asian American. Gender is not a variable if all participants in a study are cis-female.

In quantitative research, variables are defined operationally and are commonly divided into **independent variables** (active or attribute), **dependent variables**, and **extraneous variables**. Each of these topics is dealt with briefly in the following sections.

DOI: 10.4324/9781003355908-1

Operational Definitions of Variables

An operational definition describes or <u>defines a variable in terms of the operations or techniques used to make it happen or measure it</u>. When quantitative researchers describe the variables in their study, they clearly state what they mean by specifying how they measured the variable. Demographic variables like age, gender, or ethnic group are usually measured simply by asking the participant to write the correct value or choose the appropriate category from a list.

Types of treatment (or curricula) are usually operationally defined much more extensively by describing what was done during the treatment or new curriculum. Likewise, abstract concepts like mathematics knowledge, self-concept, or mathematics anxiety need to be defined operationally by spelling out in some detail how they were measured in a particular study. To do this, the investigator usually provides sample questions, appends the actual instrument, or provides a reference where more information can be found.

Independent Variables

There are two types of independent variables, **active** and **attribute**. It is important to distinguish between these types when we discuss the results of a study. As presented in more detail later, an active independent variable is a necessary but not sufficient condition to make cause and effect conclusions.

Active or manipulated independent variables. An active independent variable is a variable, such as a workshop, new curriculum, or other intervention, at least one level of which <u>is given to a group of participants, within a specified period of time during the study</u>.

For example, a researcher might investigate a new kind of therapy compared to the traditional treatment. A second example might be to study the effect of a new teaching method, such as cooperative learning, compared to independent learning. In these two examples, the variable of interest is something that is *given to* the participants. Thus, active independent variables are *given* to the participants in the study but are not necessarily given or manipulated <u>by the experimenter</u>. They may be given by a clinic, school, or someone other than the investigator, but from the participants' point of view, the situation is manipulated. To be considered an active independent variable, the treatment should be given <u>after the study is planned</u> so that there could be a pretest. Other writers have similar but, perhaps, slightly different definitions of active independent variables. **Randomized experimental** and **quasi-experimental** studies have an active independent variable.

Attribute or measured independent variables. An independent variable that cannot be manipulated, yet is a major focus of the study, can be called an attribute independent variable. In other words, the values of the independent variable are <u>preexisting attributes of the persons or their ongoing environment</u> that are not systematically changed during the study. For example, level of parental education, socioeconomic status, age, ethnic group, IQ, and self-esteem are attribute variables that could be used as attribute independent variables. Studies with only attribute independent variables are called **nonexperimental** studies.

Unlike authors of some research methods books, we do not restrict the term independent variable to those variables that are manipulated or active. We define an independent variable more broadly to include any predictors, antecedents, or *presumed* causes or influences under investigation in the study. Attributes of the participants as well as active independent variables fit within this definition. For the social sciences and education, attribute independent variables are especially important. Type of disability or level of disability may be the major focus of a study. Disability certainly qualifies as a variable because it can take on different values even though they are not *given* by the researcher during the study. For example, cerebral palsy is different from Down syndrome, which is different from spina bifida, yet all are disabilities. Also, there are different levels of the same disability. People already have defining characteristics or attributes that place them into one of two or more categories. The different disabilities are characteristics of the participants before we begin our study. Thus, we might also be interested in studying how variables that are not given or manipulated during the study, even by other persons, schools, or clinics, predict various other variables that are of interest. Similarly, age is an important variable that cannot be manipulated.

Other labels for the independent variable. SPSS uses a variety of terms, such as **factor** (Chapter 9) and **grouping variable** (Chapter 9), for the independent variables. In other cases (Chapters 7 and 8), the SPSS program and statisticians do not make a distinction between the independent and dependent variable; they just label them **variables**. For example, technically there is no independent variable for a correlation or chi-square. Even for chi-square and correlation, we think it is sometimes conceptually useful to think of one variable as the predictor (independent variable) and the other as the outcome (dependent variable); however, it is important to realize that the statistical tests of correlation and chi-square treat both variables in the same way, rather than treating one as a predictor and one as an outcome variable, as is the case in regression.

Type of independent variable and inferences about cause and effect. When we analyze data from a research study, the statistical analysis does not differentiate whether the independent variable is an active independent variable or an attribute independent variable. However, even though most statistics books use the label independent variable for both active and attribute variables, there is a crucial difference in interpretation.

A major goal of quantitative scientific research is to be able to identify a causal relationship between two variables. For those in applied disciplines, the need to demonstrate that a given intervention or treatment causes a change in behavior or performance can be extremely important. Only the approaches that have an active independent variable (randomized experimental and, to a lesser extent, quasi-experimental) can provide data that allow one to infer that the independent variable caused the change or difference in the dependent variable.

In contrast, a significant difference between or among persons with different values of an attribute independent variable should *not* lead one to conclude that the attribute independent variable caused the dependent variable to change. Thus, this distinction between active and attribute independent variables is important because terms such as **main effect** and **effect size** used by the program and most statistics books might lead one to believe that if you find a significant difference, the independent variable *caused* the difference. These terms can be misleading when the independent variable is an attribute. Of course there are other causal connections in life, such as a person pushing another person and causing them to fall. We are focusing here on quantitative research probability of one variable causing another variable to change.

Although nonexperimental studies (those with attribute independent variables) are limited in what can be said about causation, they can lead to solid conclusions about the differences between groups and about associations between variables. Furthermore, if the focus of your research is on attribute independent variables, a nonexperimental study is the *only* available approach. For example, if you are interested in learning how students varying in self-identified ethnicity differ in learning mathematical concepts, you are interested in the attribute independent variable of *self-identified ethnicity*.

Values of the independent variable. SPSS uses the term **values** to describe the several options or categories of a variable. These values are *not* necessarily ordered, and several other terms, **categories**, **levels**, **groups**, or **samples**, are sometimes used interchangeably with the term values, especially in statistics books. Suppose that an investigator is performing a study to investigate the effect of a treatment. One group of participants is assigned to the treatment group. A second group does not receive the treatment. The study could be conceptualized as having one independent variable (*treatment type*), with two values or levels (*treatment* and *no treatment*). The independent variable in this example would be classified as an active independent variable. Now, suppose instead that the investigator was interested primarily in comparing two different treatments but decided to include a third no-treatment group as a control group in the study. The study would still be conceptualized as having one active independent variable (*treatment type*), but with three values or levels (the two treatment conditions and the control condition). This variable could be diagrammed as follows:

Variable Label	*Values*	*Value Labels*
	1	= Treatment 1
Treatment type	2	= Treatment 2
	0	= No treatment (control)

As an additional example, consider *sex at birth*, which could be an attribute independent variable with two values, *male* and *female*. It could be diagrammed as follows:

Variable Label	*Values*	*Value Labels*
	0	= Male
Sex at Birth		
	1	= Female

Note that in SPSS each variable is given a **variable label**; moreover, the values, which are often categories, have **value labels** (e.g., male and female). Each value or level is assigned a number used to compute statistics. It is especially important to know the value labels when the variable is **nominal**, that is, when the values of the variable are just names and thus are not ordered (e.g., *political affiliation*, instead of a continuous variable like *height in inches* where the levels would be obvious.

Dependent Variables

The **dependent variable** <u>measures or assesses the presumed "effect" of the independent variable</u>. It is thought of as the <u>presumed outcome or criterion</u>. Dependent variables are often test scores, ratings on questionnaires, readings from instruments (e.g., electrocardiogram, galvanic skin response, etc.), or measures of physical performance. When we discuss measurement in Chapter 2, we are usually referring to the dependent variable. Dependent variables, like independent variables, must have at least two values; most of the dependent variables used in this book have <u>many values, varying from low to high</u>, for example, scores on a 100-point test, s7 they are not as easy to diagram as the independent variables shown earlier.

SPSS also uses a number of other terms for the dependent variable. **Dependent list** is used in cases where you can do the same statistic several times for a list of dependent variables (e.g., in Chapter 7 with one-way ANOVA). The term **test variable** is also used in Chapter 7 for the dependent variable in a *t* test.

Extraneous Variables

These are variables (also called nuisance variables or, in some designs, covariates) that are <u>not of interest in a particular study but could influence the dependent variable</u>. Environmental factors (e.g., temperature or distractions), time of day, and characteristics of the experimenter, teacher, or therapist are some possible extraneous variables that need to be controlled. SPSS does not use the term extraneous variable. However, sometimes such variables are "controlled" using statistics that are available in this program.

Research Hypotheses and Questions

Research hypotheses are predictive statements about the relationship between variables. **Research questions** are similar to hypotheses, except that they do not entail specific predictions and are phrased in question format. For example, one might have the following research question: "Is there a difference in students' scores on a standardized test if they took two tests in one day versus taking only one test on each of two days?" A hypothesis regarding the same issue might be: "Students who take only one test per day will score *higher* on standardized tests than will students who take two tests in one day."

We divide research questions into three broad types: difference, associational, and descriptive, as shown in the middle of Figure 1.1. The figure also shows the general and specific purposes and the general types of statistics for each of these three types of research question. We think it is educationally useful to divide inferential statistics into two types corresponding to difference and associational hypotheses or questions.[2] Difference inferential statistics (e.g., *t* test or analysis of variance) are <u>used for approaches that test for differences between groups</u>. Associational inferential statistics <u>test for associations or relationships between variables</u> and use, for example, correlation or multiple regression analysis. We utilize this contrast between difference and associational inferential statistics in Chapter 4 and later in this book.

Difference research questions. For these questions, we compare two or more different groups, each of which is composed of individuals with one of the values or levels of the independent variable. This type of question attempts to demonstrate that the <u>groups are not the same on the dependent variable</u>.

Associational research questions. Here we associate or relate two or more variables. This approach usually involves an attempt to see how two or more variables systematically covary (for example, if a person has higher values on one variable, is she or he also likely to have higher, or perhaps lower, values on another variable). An associational question could instead ask how one or more variables enable one to predict another variable.

Descriptive research questions. These are not answered with inferential statistics. They merely describe or summarize data for the sample actually studied, without trying to generalize to a larger population of individuals.

Figure 1.1 shows that <u>both difference and associational questions</u> or hypotheses <u>explore the relationships between variables</u>; however, they are conceptualized differently, as will be described shortly.[3] Note that difference and associational questions differ in specific purpose and the kinds of statistics they use to answer the question.

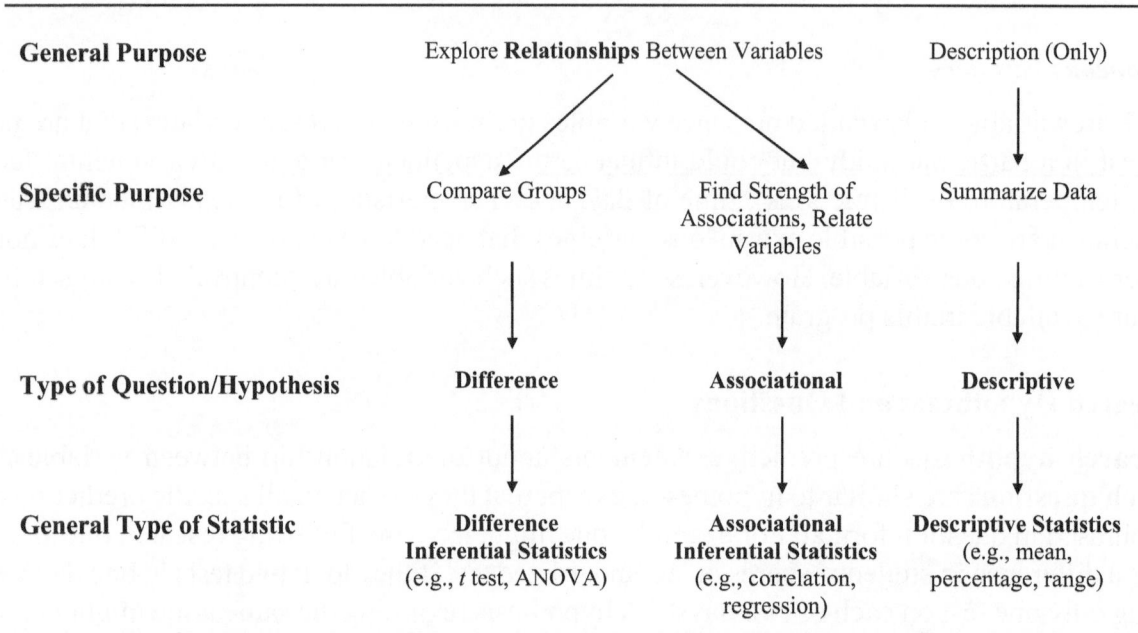

Figure 1.1 Schematic diagram showing how the purpose and type of research question correspond to the general type of statistic used in a study.

Table 1.1 provides the general format and one example of a basic difference question, a basic associational question, and a basic descriptive question. Remember that research questions are similar to hypotheses, but they are stated in question format. We think it is advisable to use the question format for the descriptive approach or when one does not have a clear directional prediction. Use the hypothesis format when you have a specific prediction based on the literature or theory. More details and examples are given in Appendix B. As implied by Figure 1.1, it is <u>acceptable to phrase any research question that involves two variables as whether or not there is a relationship between the variables</u> (e.g., is there a relationship between *political affiliation* and *math achievement* or is there a relationship between *anxiety* and *GPA*?). However, we think that phrasing questions about differences between groups as difference questions and using the relationship language for association questions only is preferable because it indicates how one is thinking about the variables and helps one identify an appropriate statistic and interpret the result.

Complex Research Questions

Some research questions involve more than two variables at a time. We call such questions and the appropriate statistics **complex**. Some of these statistics are called **multivariate** in other texts, but there is not a consistent definition of multivariate in the literature. We provide examples of

Table 1.1

Examples of Three Kinds of Basic Research Questions/Hypotheses

1. ***Basic Difference (group comparison) Questions***
   - Usually used for randomized experimental, quasi-experimental, and comparative approaches.
   - For this type of question, the groups of individuals who share a level of an active independent variable (e.g., intervention group) or an attribute independent variable (e.g., math grades—high) are compared to individuals who share the other levels of that same independent variable (e.g., control group or math grades—low) to see if the groups differ with regard to the <u>average scores</u> on the dependent variable (e.g., aggression scores).
   - Example: Do persons who experienced an emotion regulation intervention differ from those who did not experience that intervention with respect to their average aggression scores? In other words, will the average aggression score of the intervention group be significantly different from the average aggression score for the control group following the intervention?

2. ***Basic Associational (relational) Questions***
   - Used for the associational approach, in which the independent variable, if there is one, usually is continuous (i.e., has many ordered levels).
   - For this type of question, the scores on one variable (e.g., anxiety) are associated with or related to the other variable scores (e.g., GPA).
   - Example: Will students' degree of anxiety be associated with their overall GPA? In other words, will knowing students' level of anxiety tell us anything about their tendency to make higher versus lower grades? If there is a *negative* association (correlation) between anxiety scores and grade point average, those persons who have high levels of anxiety will tend to have low GPAs, those with low anxiety will tend to have high GPAs, and those in the middle on anxiety will tend to be in the middle on GPA.

3. ***Basic Descriptive Questions***
   - Used for the descriptive approach.
   - For this type of question, scores on a single variable are described in terms of their central tendency, variability, or percentages in each category/level.
   - Example: What percentage of students make a B or above? What is the average level of anxiety found in 9th grade students? The average GPA was 2.73, or 30% had high anxiety.

---

how to write certain complex research questions in Appendix B, and in Chapters 6 and 7, we introduce two complex statistics: multiple regression and factorial ANOVA. Complex statistics are discussed in more detail in our companion volume, *IBM SPSS for Intermediate Statistics*.

A Sample Research Problem: The Modified High School and Beyond (HSB) Study

The file names of the dataset used throughout this book are hsbdata.sav, which stands for high school and beyond data, and hsbdataNEW.sav, a modified version of the same dataset. They are based on a national sample of data from more than 28,000 high school students, but this dataset is a sample of 75 students drawn randomly from the larger sample. See Appendix A, Getting Started, for how to download these files to your computer. The data that we have for this sample include school outcomes such as grades and the mathematics achievement test scores of the students in high school. Also, there are several kinds of standardized test data and demographic data such as mother's and father's education. To provide an example of rating scale questionnaire data, we have included 14 items about mathematics attitudes. These data were developed for this book and thus are not really the math attitudes of the 75 students in this sample; however, they are based on real data gathered by one of the authors to study motivation. Also, we made up data for religion, ethnic group, and SAT math, which are somewhat realistic overall, and we made up the variable, academic track. These inclusions enable us to do some additional statistical analyses.

The Research Problem

Imagine that you are interested in the general problem of <u>what factors seem to influence mathematics achievement</u> at the end of high school. This general problem is the research problem. You might have some hunches or hypotheses about influences on math based on your experience and your reading of the research and popular literature. Some factors that might influence *mathematics achievement* are commonly called demographics; they include *ethnic group*, and *mother's* and *father's education*. A probable influence would be the math courses that the student has taken. We might speculate that *grades in math* and in other subjects could have an impact on *math achievement*.[4] However, other variables, such as *students' IQs* or *parents' encouragement*, could be the actual causes of both high grades and math achievement. Such variables could influence what courses one took and the grades one received, and they might be correlates of the demographic variables. We might wonder how spatial performance scores, such as pattern or *mosaic test scores* and *visualization scores*, might enable a more complete understanding of the problem, and whether these skills seem to be influenced by the same factors as is *math achievement*.

The HSB Variables[5]

Before we state the research questions in more formal ways, we need to step back and discuss the types of variables and the approaches that might be used to study the previous research problem. We need to identify the independent/antecedent (presumed causes) variables, the dependent/outcome variable(s), and any extraneous variables.

The primary dependent variable. Because the research problem focuses on achievement tests at the end of the senior year, the primary dependent variable (or **Target**) is *math achievement*.

Independent and extraneous variables. *Father's* and *mother's education* and participant's *ethnicity, religion*, and *academic track* are best considered to be **input** (the SPSS term), antecedent, or independent variables in this study. These variables would usually be thought of as independent rather than as dependent variables because they occurred before the math achievement test and do not vary during the study. However, some of these variables, such as ethnicity and religion, might be viewed as extraneous variables that need to be "controlled."

Many of the variables, including *visualization* and *mosaic pattern scores*, could be viewed *either* as independent or dependent variables depending on the specific research question because they were measured at approximately the same time as math achievement. We have labeled them **Both** under **Role**. Note that student's class is a constant and is not a variable in this study because all the participants are high school seniors (i.e., it does not vary; it is the population of interest).

Types of independent variables. As we previously discussed, independent variables can be **active** (given to the participant during the study or manipulated by the investigator) or **attributes** of the participants or their environments. Are there any **active** independent variables in this study? No! There is no intervention, new curriculum, or similar treatment. All the independent variables, then, are attribute variables because they are attributes or characteristics of these high school students. <u>Given that all the independent variables are attributes, the research approach cannot be experimental.</u> This means that we will *not* be able to draw definite conclusions about cause and effect (i.e., we will find out what is related to math achievement, but <u>we will not know for sure what causes or influences</u> math achievement).

Now we examine the *hsbdata.sav* file that you will use to study this complex research problem. We have provided on the website described in the Preface these data for each of the 75 participants on 38 variables. The variables in the *hsbdata.sav* file have already been labeled (see Figure 1.2) and entered (see Figure 1.3) to enable you to get started on analyses quickly. This file contains data files for you to use, but it <u>does not include the actual SPSS Statistics program</u> to which you will need access in order to do the problems.

The Variable View

Figure 1.2 is a piece of what is called the **Variable View** in the **Data Editor** for the *hsbdata.sav* file. It also shows information about each of the first 30 variables. When you open this file and click on **Variable View** at the bottom left corner of the screen, this is what you will see. We describe what is in the variable view screen in more detail in Chapter 2; for now, focus on the Name, Label, Values, and Missing columns. **Name** is a short name for each variable (e.g., *faed* or *alg1*).[6] **Label** is a longer label for the variable (e.g., *father's education* or *algebra 1 in h.s.*). The **Values** column contains the **value labels**, but you can see only the label for one value at a time (e.g., calculus in high school 0 = not taken). That is, you cannot see that 1 = taken unless you click on the gray square in the value column. The **Missing** column indicates whether there are any special, user-identified missing values. **None** just means that there <u>are no special missing values</u>, just the usual **system missing** value, which is a blank. Appendix A shows how to get this data.

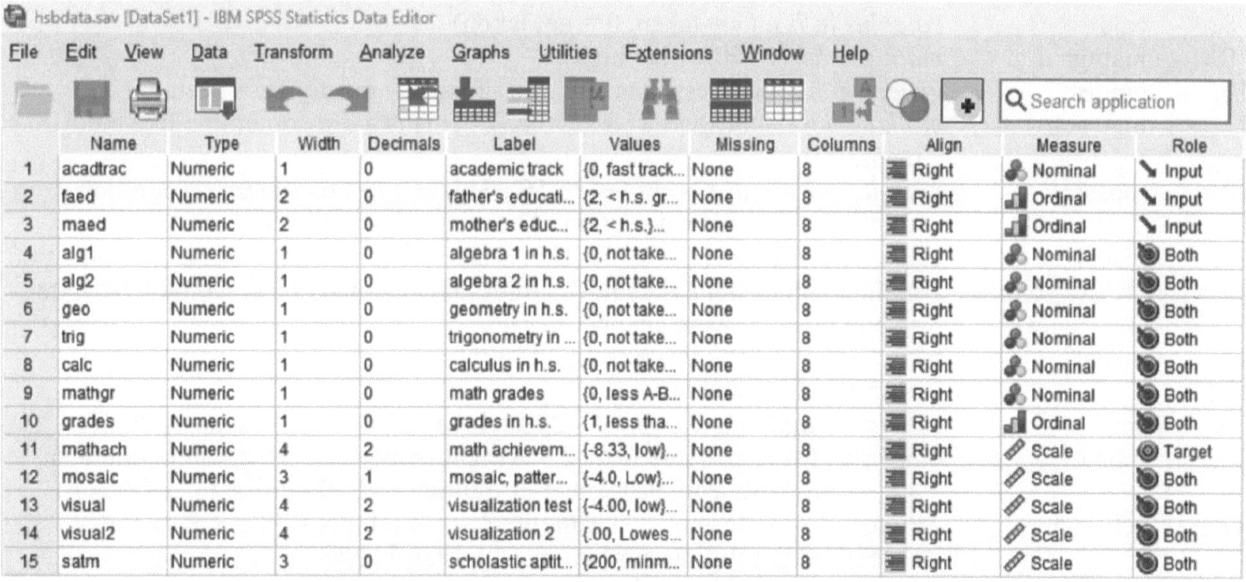

hsbdata.sav [DataSet1] – IBM SPSS Statistics Data Editor

File   Edit   View   Data   Transform   Analyze   Graphs   Utilities   Extensions   Window   Help

	Name	Type	Width	Decimals	Label	Values	Missing	Columns	Align	Measure	Role
1	acadtrac	Numeric	1	0	academic track	{0, fast track...	None	8	Right	Nominal	Input
2	faed	Numeric	2	0	father's educati...	{2, < h.s. gr...	None	8	Right	Ordinal	Input
3	maed	Numeric	2	0	mother's educ...	{2, < h.s.}...	None	8	Right	Ordinal	Input
4	alg1	Numeric	1	0	algebra 1 in h.s.	{0, not take...	None	8	Right	Nominal	Both
5	alg2	Numeric	1	0	algebra 2 in h.s.	{0, not take...	None	8	Right	Nominal	Both
6	geo	Numeric	1	0	geometry in h.s.	{0, not take...	None	8	Right	Nominal	Both
7	trig	Numeric	1	0	trigonometry in ...	{0, not take...	None	8	Right	Nominal	Both
8	calc	Numeric	1	0	calculus in h.s.	{0, not take...	None	8	Right	Nominal	Both
9	mathgr	Numeric	1	0	math grades	{0, less A-B...	None	8	Right	Nominal	Both
10	grades	Numeric	1	0	grades in h.s.	{1, less tha...	None	8	Right	Ordinal	Both
11	mathach	Numeric	4	2	math achievem...	{-8.33, low}...	None	8	Right	Scale	Target
12	mosaic	Numeric	3	1	mosaic, patter...	{-4.0, Low}...	None	8	Right	Scale	Both
13	visual	Numeric	4	2	visualization test	{-4.00, low}...	None	8	Right	Scale	Both
14	visual2	Numeric	4	2	visualization 2	{.00, Lowes...	None	8	Right	Scale	Both
15	satm	Numeric	3	0	scholastic aptit...	{200, minm...	None	8	Right	Scale	Both

Figure 1.2 Part of the hsbdata.sav Variable View in the data editor.

Variables in the Modified HSB Dataset

The 39 variables shown in Table 1.2 (with the values/levels or range of their values in parentheses) are found in the *hsbdata.sav* file. Also included in Table 1.2, for completeness, are seven variables (numbers 40–46) that are not yet in the *hsbdata.sav* dataset because you will compute them in Chapter 2. Note that variables 34–39 have been computed already from the math attitude variables (19–32) so that you would have fewer new variables to compute in Chapter 2.

The variables of *ethnic* and *religion* were added to the dataset to provide true nominal (unordered) variables with a few (4 and 3) levels or values. In addition, for *ethnic* and *religion*, we have made two missing value codes to illustrate this possibility. All other variables use blanks, the system missing value, for missing data.

For *ethnicity*, 98 indicates multiethnic and other. For *religion*, all the high school students who were not *protestant* or *catholic* or said they were *not religious* were coded 98 and considered to be missing because none of the other religious affiliations (e.g., Muslim) had enough members to make a reasonable size group. Those who left the ethnicity or religion questions blank were coded as 99, also missing.

Table 1.2

HSB Variable Descriptions

	Name	Label (and Values)
		Demographic School and Test Variables
1.	acadtrac	*academic track* (0 = fast track, 1 = regular track)
2.	faed	*father's education* (2 = less than h.s. grad to 10 = PhD/MD)
3.	maed	*mother's education* (2 = less than h.s. grad to 10 = PhD/MD)
4.	alg1	*algebra 1 in h.s.* (1 = taken, 0 = not taken)
5.	alg2	*algebra 2 in h.s.* (1 = taken, 0 = not taken)
6.	geo	*geometry in h.s.* (1 = taken, 0 = not taken)
7.	trig	*trigonometry in h.s.* (1 = taken, 0 = not taken)
8.	calc	*calculus in h.s.* (1 = taken, 0 = not taken)
9.	mathgr	*math grades* (0 = low, 1 = high)
10.	grades	*grades in h.s.* (1 = less than a D average to 8 = mostly an A average)
11.	mathach	*math achievement score* (−8.33 to 25).[7] This is a test something like the ACT math.
12.	mosaic	*mosaic, pattern test score* (−4 to 56). This is a test of pattern recognition ability involving the detection of relationships in patterns of tiles.
13.	visual	*visualization score* (−4 to 16). This is a 16-item test that assesses visualization in three dimensions (i.e., how a three-dimensional object would look if its spatial position were changed).
14.	visual2	*visualization 2.* The visualization test rated by a second researcher who observed the same test.
15.	satm	*scholastic aptitude test—math* (200 = lowest, 800 = highest possible)
16.	ethnic	*ethnicity* (1 = Euro-American, 2 = African-American, 3 = Latino-American, 4 = Asian-American, 98 = other or multiethnic, 99 = missing, left blank)
17.	religion	*religion* (1 = protestant, 2 = catholic, 3 = not religious, 98 = chose one of several other religious affiliations, 99 = left blank)
18.	ethnic2	*ethnicity reported by student* (same as values for ethnic)
		Math Attitude Questions 1–14 (Rated from 1 = very atypical to 4 = very typical)
19.	item01	*Motivation* "I practice math skills until I can do them well."
20.	item02	*Pleasure* "I feel happy after solving a hard problem."
21.	item03	*Competence* "I solve math problems quickly."
22.	item04	*(low) motiv* "I give up easily instead of persisting if a math problem is difficult."
23.	item05	*(low) comp* "I am a little slow catching on to new topics in math."
24.	item06	*(low) pleas* "I do not get much pleasure out of math problems."

(*Continued*)

Table 1.2 (Continued)

	Name	Label (and Values)
25.	item07	*Motivation* "I prefer to figure out how to solve problems without asking for help."
26.	item08	*(low) motiv* "I do not keep at it very long when a math problem is challenging."
27.	item09	*Competence* "I am very competent at math."
28.	item10	*(low) pleas* "I smile only a little (or not at all) when I solve a math problem."
29.	item11	*(low) comp* "I have some difficulties doing math as well as other kids my age."
30.	item12	*Motivation* "I try to complete my math problems even if it takes a long time to finish."
31.	item13	*Motivation* "I explore all possible solutions of a complex problem before going on to another one."
32.	item14	*Pleasure* "I really enjoy doing math problems."
33.	mosaic2	*Mosaic pattern test 2.* The score rated by a second researcher who observed the same test.

New Variables Computed From the Previous Variables

34.	item04r	*item04 reversed* (4 now = high motivation)
35.	item05r	*item05 reversed* (4 now = high competence)
36.	item08r	*item08 reversed* (4 now = high motivation)
37.	item11r	*item11 reversed* (4 now = high competence)
38.	competence	*competence scale.* An average computed as follows: (item03 + item05r + item09 + item11r)/4
39.	motivation	*motivation scale* (item01 + item04r + item07 + item08r + item12 + item13)/6

Variables to be Computed in Chapter 2

40.	mathcrs	*math courses taken* (0 = none, 5 = all five)
41.	faedRevis	*father's educ revised* (1 = HS grad or less, 2 = some college, 3 = BS or more)
42.	maedRevis	*mother's educ revised* (1 = HS grad or less, 2 = some college, 3 = BS or more)
43.	item06r	*item06 reversed* (4 now = high pleasure)
44.	item10r	*item10 reversed* (4 now = high pleasure)
45.	pleasure	*pleasure scale* (item02 + item06r + item10r + item14)/4
46.	parEduc	*parents' education* (average of the <u>unrevised</u> mother's and father's educations)

The Raw HSB Data and Data Editor

Figure 1.3 is a piece of the *hsbdata.sav* file showing raw data for the first 34 student participants for variables 1 through 22 (academic track through item04). When you open this file and click on **Data View** at the bottom left corner of the screen, this is what you will see. Notice the short variable names (e.g., *faed, maed, calc,* etc.) at the top of the hsbdata file. Be aware that the participants are listed down the left side of the page, and the variables are listed across the top. For all the statistics in <u>this</u> book, <u>you will always enter data this way</u>. If a variable is measured more than once, such as *visual* and *visual2* (see Figure 1.3), it will be entered as two variables with slightly different names.

Note that in Figure 1.3, most of the values are single digits, but *mathach, mosaic,* and *visual* include some decimals and even negative numbers. Notice also that some cells, like *father's education* for participant 5, are blank because a datum is missing. Perhaps Participant 5 did not know her father's education. Blank is the system missing value that can be used for any missing data in an SPSS data file. We suggest that you leave missing data blank unless there is some reason you

NOTE.[8]

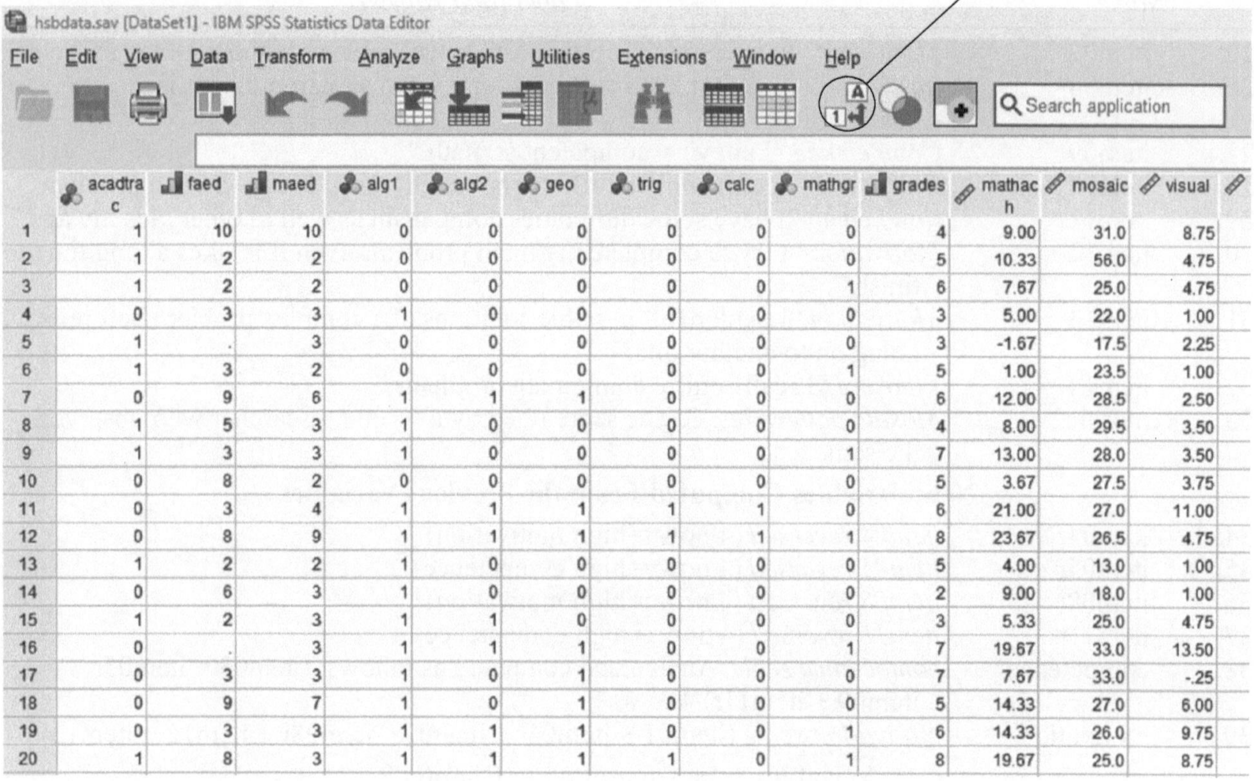

	acadtrac	faed	maed	alg1	alg2	geo	trig	calc	mathgr	grades	mathach	mosaic	visual
1	1	10	10	0	0	0	0	0	0	4	9.00	31.0	8.75
2	1	2	2	0	0	0	0	0	0	5	10.33	56.0	4.75
3	1	2	2	0	0	0	0	0	1	6	7.67	25.0	4.75
4	0	3	3	1	0	0	0	0	0	3	5.00	22.0	1.00
5	1	.	3	0	0	0	0	0	0	3	-1.67	17.5	2.25
6	1	3	2	0	0	0	0	0	1	5	1.00	23.5	1.00
7	0	9	6	1	1	1	0	0	0	6	12.00	28.5	2.50
8	1	5	3	1	0	0	0	0	0	4	8.00	29.5	3.50
9	1	3	3	1	0	0	0	0	1	7	13.00	28.0	3.50
10	0	8	2	0	0	0	0	0	0	5	3.67	27.5	3.75
11	0	3	4	1	1	1	1	1	0	6	21.00	27.0	11.00
12	0	8	9	1	1	1	1	1	1	8	23.67	26.5	4.75
13	1	2	2	1	0	0	0	0	0	5	4.00	13.0	1.00
14	0	6	3	1	0	0	0	0	0	2	9.00	18.0	1.00
15	1	2	3	1	1	0	0	0	0	3	5.33	25.0	4.75
16	0	.	3	1	1	1	0	0	1	7	19.67	33.0	13.50
17	0	3	3	1	0	1	0	0	0	7	7.67	33.0	-.25
18	0	9	7	1	0	1	0	0	0	5	14.33	27.0	6.00
19	0	3	3	1	1	1	0	0	1	6	14.33	26.0	9.75
20	1	8	3	1	1	1	1	0	1	8	19.67	25.0	8.75

Figure 1.3 Part of the hsbdata Data View in the data editor.

need to distinguish among types of or reasons for missing data (see *religion* for subjects 9, 11, and 12 and the description of what 98 and 99 mean in Table 1.2).

Note also that there is a column headed by "**Measure**". This indicates the level of measurement that is assumed for each variable. Level of measurement is an important attribute of the variable that has important implications for which statistical analyses should be conducted with the variable. We will discuss this concept now.

Levels of Measurement

Measurement is the assignment of numbers or symbols to the different characteristics (values) of variables according to rules. In order to understand your variables, it is important to know their level of measurement. Depending on the level of measurement of a variable, the data can mean different things. For example, the number 2 might indicate a score of two; it might indicate that the subject was ranked second in the class; it might even indicate that the subject was in the treatment group. To help understand these differences, types or levels of variables have been identified. It is common and traditional to discuss four levels or scales of measurement: **nominal**, **ordinal**, **interval**, and **ratio**, which vary from the unordered (nominal) to the highest level (ratio).[9] These four traditional terms are not the same as those used in the SPSS program, and we think that they are not always the most useful for determining what statistics to use.

SPSS uses three terms (**nominal**, **ordinal**, and **scale**) for the levels or types of measurement. How these correspond to the traditional terms is shown in Table 1.3. When you name and label variables with the SPSS program, you have the opportunity to select one of these three types of measurement, as was demonstrated in Chapter 2 (see Figure 2.2). An appropriate choice indicates that you understand your data, and it may help guide your selection of statistics.

We believe that the terms **nominal, dichotomous, ordinal**, and **approximately normal** (for normally distributed) are usually more useful than the traditional or SPSS measurement terms for the selection and interpretation of statistics. In part, this is because statisticians disagree about the usefulness of the traditional levels of measurement in determining the appropriate selection of statistics. Furthermore, our experience is that the traditional terms are frequently misunderstood and applied inappropriately. The main problem with the SPSS terms is that the term scale is not commonly used as a measurement level, and it has other meanings (see endnote 2) that make its use here confusing. Hopefully, our terms are clear and useful.

Table 1.3 compares the three sets of terms and provides a summary description of our definitions of them. Professors differ in the terminology they prefer and on how much importance to place on levels or scales of measurement. Unfortunately, you will see all of these terms and others mentioned in textbooks and articles.

Nominal Variables

This is the most basic or lowest level of measurement, in which the numerals assigned to each category stand for the <u>name</u> of the category, but they have no implied order or value. For example, in the HSB study, the values for the *religion* variable are 1 = *protestant*, 2 = *catholic*, 3 = *not religious*. This does not mean that two Protestants equal one Catholic, or any other typical mathematical uses of the numerals. The same reasoning applies to many other true nominal variables, such as ethnic group, type of disability, or section number in a class schedule. In each of these cases, the categories are distinct, non-overlapping, and not ordered. Each category or group in the modified HSB variable *ethnicity* is different from every other, but there is no order to the categories. Thus, the categories could have been numbered 1 for *Asian American*, 2 for *Latino American*, 3 for *African American*, and 4 for *European American*, the reverse, or any combination of assigning one number to each category.

What this implies is that you must *not* treat the numbers used for identifying nominal categories as if they were numbers that could be used in a formula, added together, subtracted from one another, or used to compute an average. Average ethnic group makes no sense. However, if you ask SPSS to compute the average ethnic group, it will do so and give you meaningless information. The important aspect of nominal measurement is to have clearly defined, non-overlapping, and mutually exclusive categories that can be coded reliably by observers or by self-report.

Table 1.3

Comparison of Traditional, SPSS, and Our Measurement Terms

Traditional Term	Traditional Definition	SPSS Term	Our Term	Our Definitions
Nominal	Two or more <u>unordered</u> categories.	Nominal	Nominal	Three or more <u>unordered</u> categories.
NA	NA	NA	Dichotomous	Two categories, either ordered or unordered.

(*Continued*)

Table 1.3 (Continued)

Traditional Term	Traditional Definition	SPSS Term	Our Term	Our Definitions
Ordinal	Ordered levels, in which the difference in magnitude between levels is not equal (e.g., scores such as 1 and 2, or 2 and 3) and distorts the meaning of the data.	Ordinal	Ordinal	Three or more ordered levels, in which the difference in magnitude between pairs of adjacent levels is not equal. If the frequency distribution of the scores is not normally distributed (often it is skewed) label it ordinal.
Interval & Ratio	**Interval:** ordered levels, in which the difference between levels is equal, but no true zero. **Ratio:** ordered levels; the difference between levels is equal, and there is a true zero.	Scale	Approximately Normal (or Normal)	Many (at least five) ordered levels or scores, with the frequency distribution of the scores being approximately normal.

Using only nominal variables does dramatically reduce the statistics that can be used with your data, but it does not altogether eliminate the possible use of statistics to summarize your data and make inferences. Therefore, even when the data are unordered or nominal categories, your research may benefit from the use of appropriate statistics. Later we discuss the types of statistics, both descriptive and inferential, that are appropriate for nominal data.

Other terms for nominal variables. Unfortunately, the literature is full of similar but not identical terms to describe the measurement aspects of variables. **Categorical**, **qualitative**, and **discrete** are terms sometimes used interchangeably with nominal, but we think that nominal is better because it is possible to have ordered, discrete categories (e.g., low, medium, and high IQ, which we and other researchers would consider an ordinal variable). "Qualitative" is also used to discuss a different approach to doing research, with important differences in philosophy, assumptions, and methods of conducting research.

Dichotomous Variables

Dichotomous variables always have only two levels or categories. In some cases, they may have an implied order (e.g., *math grades* in high school are coded 0 for *less than an A or B* average and 1 for *mostly A or B*). Other dichotomous variables do not have any order to the categories (e.g., *fast track* or *regular track*). For many purposes, it is best to use the same statistics for dichotomous and nominal variables. However, a statistic such as the mean or average, which would be meaningless for a three or more category nominal variable (e.g., *ethnicity*), does have meaning when there are only two categories, especially when these levels are coded as 0 and 1 (a dummy variable). For example, in the HSB data, the average *academic track* is .55 (with *fast track* = 0 and *regular track* = 1). This means that 55% of the participants were *on the regular track*, the level coded as 1. Furthermore, we see with multiple regression that dichotomous variables, usually coded as *dummy variables*, can be used as independent variables along with other variables that are normally distributed.

Other terms for dichotomous variables. In the **Variable View**, we label dichotomous variables "nominal," and this is common in textbooks. However, please remember that dichotomous variables are really a special case and for some purposes they can be treated as if they were normal or scale. Dichotomous data have two discrete categories and are sometimes called **discrete variables**, **categorical variables**, or **dummy variables**.

Ordinal Variables

In ordinal measurement, there are not only mutually exclusive categories as in nominal scales, but the categories are ordered from low to high, such that ranks could be assigned (e.g., 1st, 2nd, 3rd). Thus, in an ordinal scale, higher numbers tell you that the person is rated higher or more of something, but the intervals between the various categories are not equal. Our classification of a variable as ordinal focuses on whether the frequency counts for each category or value are distributed like the bell-shaped, normal curve with more responses in the middle categories and fewer in the lowest and highest categories. If not approximately normal, we would call the variable ordinal. Ordered variables with only a few categories (say 2–4) would also be called ordinal. As indicated in Table 1.3, however, the traditional definition of ordinal focuses on whether the differences between pairs of levels are equal. This can be important, for example, if one will be creating summed or averaged scores (as in subscales of a questionnaire that involve aggregating a set of questionnaire items). If differences between levels are *meaningfully* unequal, then averaging a score of 5 (e.g., indicating the participants' age is 65+) and a score of 2 (e.g., indicating that the participants' age is 20–25) may not make sense. Averaging the *ranks* of the scores may be more meaningful if it is clear that they are ordered but that the differences between adjacent scores differ across levels of the variable. However, sometimes even if the differences between levels are not literally equal (e.g., the difference between a level indicating infancy and a level indicating preschool is not equal in years to the difference between a level of "young adulthood" and "older adulthood"), it may be reasonable to treat the levels as interval level data if the levels comprise the most meaningful distinctions and data are normally distributed.

Other terms for ordinal variables. Some authors use the term **ranks** interchangeably with ordinal. However, most analyses that are designed for use with ordinal data (nonparametric tests) rank the data as a part of the procedure, assuming that the data you are entering are not already ranked. Moreover, the process of ranking changes the distribution of data such that it can be used in many analyses usually requiring normally distributed data. Ordinal data are often **categorical** (e.g., good, better, best are three ordered *categories*), so categorical is sometimes used to include both nominal and ordinal data. The categories may be **discrete** (e.g., number of children in a family is a discrete number; e.g., 1 or 2, etc.; it does not make sense for one family to have a number of children in between 1 and 2).

Approximately Normal (or Scale) Variables[10]

Approximately normally distributed variables not only have levels or scores that are *ordered* from low to high, but also, as stated in Table 1.3, the frequencies of the scores are approximately normally distributed. That is, most scores are somewhere in the middle with similar smaller numbers of low and high scores. Thus, a Likert scale, such as strongly agree to strongly disagree, would be considered normal if the frequency distribution was approximately normal. We think normality, because it is an assumption of many statistics, should be the focus of this highest level of measurement. Many normal variables are continuous (i.e., they have an infinite number of possible values within some range). If not continuous, we suggest that there be at least five ordered values or levels and that they have an implicit, underlying continuous nature. For example, a 5-point Likert scale has only five response categories, but in theory, a person's rating could fall anywhere between 1 and 5 (e.g., halfway between 3 and 4).

Other terms for approximately normal variables. Some terms that you will see in the literature for variables that vary from low to high and are assumed to be normally distributed are **continuous**, **dimensional**, and **quantitative**. SPSS Statistics uses **scale**, as previously noted. Traditional measurement terminology uses the terms interval and ratio. Because they are common in the literature and overlapping with the SPSS term scale, we describe them briefly. **Interval** variables have ordered categories that are equally spaced (i.e., have equal intervals between them). Most physical measurements (e.g., *length, weight, temperature*, etc.) have equal intervals between them. Some physical measurements (e.g., *temperature in degrees Kelvin or amount of sugar in a beverage*) in fact not only have equal intervals between the levels or scores, but also a true zero, which means, in the previous examples, no movement of atoms or no sugar. Such variables are called **ratio** variables. It is rare for a psychological scale to have a true zero and thus even if they are very well-constructed equal interval scales, it is not possible to say that zero means that one has no intelligence, no extroversion, or no attitude of a certain type. However, the differences between interval and ratio scales are not important for us because we can do all of the types of statistics that we have available with interval data. SPSS Statistics terminology supports this non-distinction by using the term **scale** for both interval and ratio data. An assumption of most parametric statistics <u>is that the variables be approximately normally distributed, not whether they have equal intervals between levels or a true zero</u>.

Labeling Levels of Measurement in SPSS Statistics

When you label variables with this program, the **Measure** column provides only three choices: nominal, ordinal, or scale. How do you decide which one to use?

Labeling variables as nominal. If the variable has only two levels (e.g., Yes or No, Pass or Fail), most researchers and we would label it **nominal** in the SPSS Statistics **Variable View** because that is traditional, and it is often best to use the same statistics with dichotomous variables that you would with a nominal variable. As mentioned earlier, there are times when dichotomous variables can be treated as if they were ordinal; however, as long as you use numbers to code them, SPSS will still allow you to use them in such analyses. If there are three or more categories or values, you need to determine whether the categories are ordered (vary from low to high) or not. If the categories are just different names and not ordered, label the variable as **nominal** in the Variable View. Especially if there are more than two categories, this <u>distinction between nominal and ordered variables makes a lot of difference</u> in choosing and interpreting appropriate statistics.

Labeling variables as ordinal. If the categories or values of a variable vary from low to high (i.e., are ordered), and there are only three or four such values (e.g., good, better, best, or strongly disagree, disagree, agree, strongly agree), we recommend that you initially label the variable **ordinal**. Also, even if there are five or more ordered levels or values of a variable, <u>if</u> you suspect that the frequency distribution of the variable is substantially non-normal, label the variable

ordinal. That is, if you <u>do not</u> think that the distribution is approximately symmetrical and that most of the participants had scores somewhere in the middle of the distribution, call the variable ordinal. If most of the participants are thought to be either quite high or low or you suspect that the distribution will be some shape other than bell-shaped, label the variable ordinal.

Labeling variables as scale. If the variable has five or more <u>ordered</u> categories or values and you have no reason to suspect that the distribution is non-normal, label the variable **scale** in the variable view **Measure** column. If the variable is essentially continuous (e.g., is measured to one or more decimal places or is the average of several items), it is likely to be at least approximately normally distributed, so call it **scale**. As you will see in Chapter 2, we recommend that you check the skewness of your variables with five or more ordered levels and then adjust what you initially called a variable's measurement, if necessary. That is, you might want to change it from ordinal to scale if it turns out to be approximately normal or change from scale to ordinal if it turns out to be too skewed.[11]

Why We Prefer Our Four Levels of Measurement: A Review

As shown in Tables 1.3 and 1.4, we distinguish between four levels of measurement: nominal, dichotomous, ordinal, and normal. Even though you cannot label variables as dichotomous or normal in the SPSS Statistics variable view, we think that these four levels are conceptually and practically useful. Remember that because dichotomous variables form a special case, they can be used and interpreted much like normally distributed variables, which is why we think it is good to distinguish between nominal and dichotomous even though the SPSS program does not.

Likewise, we think that normally distributed or normal is a better label than the term scale because the latter could easily be confused with other uses of the term scale (see endnote 2) and because whether or not the variable is approximately normally distributed is, for us, what distinguishes it from an ordinal variable. Furthermore, what is important for most of the inferential statistics (e.g., *t* test) that you will compute with SPSS is the assumption that the dependent variable must be at least approximately normally distributed.

Remember that in SPSS, there are only three measurement types or levels, and you are the one who determines if the variable is called nominal, ordinal, or scale. We called dichotomous variables nominal, and we have labeled approximately normal variables as scale in our hsbdata file.

Table 1.4

Characteristics and Examples of Our Four Levels of Measurement

	Nominal	*Dichotomous*	*Ordinal*	*Normal (scale)*
Characteristics	3+ levels Not ordered True categories Names, labels	2 levels Ordered or not	3+ levels Ordered levels Unequal intervals between levels Not normally distributed, often skewed	5+ levels Ordered levels Approximately normally distributed Equal intervals between levels
Examples	Ethnicity Religion Curriculum type Hair color	Experimental vs. Control Math grades (high vs. low)	Competence scale Mother's education	SAT math Math achievement Height

Research Questions for the Modified HSB Study[12]

In this book, we generate a large number of research questions from the modified HSB dataset. In this section, we list some research questions, which the HSB data will help answer, in order to give you an idea of the range of types of questions that one might have in a typical research project like a thesis or dissertation. In addition to the **difference** and **associational questions** that are commonly seen in a research article, we have asked **descriptive questions** and questions about assumptions in the early assignments. Templates for writing the research problem and research questions or hypotheses are given in Appendix B, which should help you write questions for your own research.

1. Often, we start with basic **descriptive questions** about the demographics of the sample. Results of these analyses typically would be written up in the Method section, rather than the Results section. Thus, we could answer, with the methods discussed in Chapter 2, the following basic descriptive question: "What is the average educational level of the fathers of the students in this sample?" "What percentage of the students has taken calculus?"

2. In the assignments for Chapter 2, we also examine whether the continuous variables (those that might be used to answer associational questions) are distributed normally, an **assumption** of many statistics. One question is "Are the frequency distributions of the math achievement scores markedly skewed, that is, different from the normal curve distribution?"

3. Chapter 4 presents several statistics that are frequently used to check the data for reliability and validity and to reduce a large number of variables, such as the math attitude questions, to a smaller, more manageable number of composite variables. One usually wouldn't formally state research questions about reliability and validity in a research article. However, this chapter could help you answer questions such as "Are the codes for ethnicity from the school records and the students similar, i.e., reliable?"

4. Tables cross-tabulating two categorical variables (ones with a few values or categories) are computed in Chapter 5. Cross-tabulation and the chi-square statistic can answer research questions such as "Is there a relationship between calc (calculus taken or not) and math grades (high or low)?"

5. In Chapter 6, we answer **basic associational research questions** (using Pearson product-moment correlation coefficients) such as "Is there a positive association/relationship between grades in high school and math achievement?" This assignment also produces a correlation matrix of all the correlations among several key variables, including math achievement. Similar matrixes will provide the basis for computing multiple regression. In Chapter 4, correlation is also used to assess reliability.

6. Chapter 6 also poses a **complex associational question** such as "How well does a combination of variables predict math achievement?" in order to introduce you to multiple regression.

7. In Chapter 7, we ask several basic **difference questions** comparing two groups, utilizing an independent samples *t* test. For example, we ask, "Do those who have had geometry have higher visualization scores than those who did not have geometry?" In Chapter 7, we also ask basic difference questions in which the independent variable has three or more values. For example, "Are there differences among the three fathers' education groups in regard to average scores on math achievement?" An answer is based on a one-way or single factor analysis of variance (ANOVA).

8. **Complex difference questions** are also asked in Chapter 7. One *set* of three questions is as follows: (1) "Is there a difference between students who have fathers with no college, some

college, or a BS or more with respect to the student's math achievement?" (2) "Is there a difference between students who had a B or better math grade average and those with less than a B average on a math achievement test at the end of high school?" and (3) "Is there an interaction between a father's education and math grades with respect to math achievement?" Answers to this set of three questions are based on factorial ANOVA, introduced briefly in Chapter 7.

This introduction to the research problem and questions raised by the HSB dataset should help make the assignments meaningful, and it should provide a guide and some examples for your own research.

Interpretation Questions

1.1. Compare the terms *active independent variable* and *attribute independent variable*. What are the similarities and differences?

1.2. What kind of independent variable (active or attribute) is necessary to infer cause? Can one *always* infer cause from this type of independent variable? If so, why? If not, when can one infer cause, and when might causal inferences be more questionable?

1.3. What is the difference between the independent variable and the dependent variable?

1.4. Compare and contrast associational, difference, and descriptive types of research questions.

1.5. Write a research question *and* a corresponding hypothesis regarding variables of interest to you but not in the HSB dataset. Is it an associational, difference, or descriptive question?

1.6. Using one or more of the following HSB variables, *religion*, *mosaic pattern test*, and *visualization score*:

(a) Write an associational question.
(b) Write a difference question.
(c) Write a descriptive question.

1.7. Why is it important to distinguish dichotomous variables from nominal variables with more than two categories?

Notes

1 To help you we have identified the variable names, labels, and values using italics (e.g., *age* and *female*) and have put in bold the terms used in the SPSS screens and outputs (e.g., **Data Editor**). We also use bold for other key terms when they are introduced, defined, or are important to understanding. Underlines are used to focus your attention on critical points or phrases that could be missed. Italics are occasionally used, as is commonly the case, for emphasizing words and for the titles of books.

2 We realize that all parametric inferential statistics are relational so this dichotomy of using one type of data analysis procedure to test for differences (when there are a few values or levels of the independent variables) and another type of data analysis procedure to test for associations (when there are continuous independent variables) is somewhat artificial. Both continuous and categorical independent variables can be used in a general linear model approach to data analysis. However, we think that the distinction is useful because most researchers utilize the dichotomy in selecting statistics for data analysis.

3 This similarity is in agreement with the statement by statisticians that all common parametric inferential statistics are relational. We use the term associational for the second type of research question

rather than relational or correlational to distinguish it from the *general purpose* of both difference and associational questions/hypotheses, which is to study relationships. Also we want to distinguish between correlation, as a specific statistical technique, and the broader type of associational question and that group of statistics.

4   We have decided to use the short version of mathematics (i.e., math) throughout the book to save space and because it is used in common language.

5   New to version 18 of the program was the Role column in the Variable View. SPSS now allows the user to assign the term Target to dependent variables, Input for independent variables, and Both for variables that are used as either or both independent and dependent variables.

6   The variable **Name** must start with a letter and must <u>not</u> contain <u>blank spaces</u> or certain special characters (e.g., !, ?, ', or *). Certain reserved keywords cannot be used as variable names (e.g., ALL, AND, EQ, BY, TO, or WITH). The variable **label** can be up to 40 characters including spaces, but the outputs are neater if you keep labels to 20 characters or less.

7   Negative test scores may result from a penalty for guessing.

8   If the values for *calc* are shown as *not taken* or *taken*, the **value labels** rather than the numerals are being displayed. In that case, click on the circled symbol to change the format to show only the numeric values for each variable.

9   Unfortunately, the terms "level" and "scale" are used several ways in research. Levels refer to the categories or values of a variable (e.g., experimental or control group or 1, 2, or 3); level can also refer to the three or four different types of measurement (nominal, ordinal, etc.). These several types of measurement have also been called "scales of measurement," but SPSS uses scale specifically for the highest type or level of measurement. Other researchers use scale to describe questionnaire items that are rated from strongly disagree to strongly agree (Likert scale) and for the sum of such items (summated scale). We wish there were not so many uses of these terms; the best we can do is try to be clear about our usage.

10 Some students confuse the terms Nominal and Normal because they both start with "No" and are about the same length. They mean very different things so it is important not to confuse them.

11 Another alternative would be to transform the variable to normalize the distribution.

12 The High School and Beyond (HSB) study was conducted by the National Opinion Research Center. The example discussed here and throughout the book is based on 13 variables obtained from a random sample of 75 out of 28,240 high school seniors. These variables include achievement scores, grades, and demographics. The raw data for the 13 variables were slightly modified from published HSB data. That file had no missing data, which is unusual in behavioral science research.

2   Getting Data Ready for Analysis and Understanding It

Data Collection, Coding, and Description

In this chapter, we discuss collecting, coding, and cleaning up your data prior to conducting any of your main analyses. Cleaning data is extremely important as errors, inappropriately coded data (e.g., letters instead of numbers), or other issues need to be corrected before conducting any statistics. This chapter focuses on the first steps to take to utilize and understand your data and includes some of the most important steps that you should be sure to do with all your data files.

Plan the Study, Pilot Test, and Collect Data

Select or develop the instrument(s). Before you collect data, it is important to identify reliable and valid methods of collecting data that answer your research questions or hypotheses. Most research methods books discuss this part of the research process extensively (e.g., see Gliner et al., 2017). If there is an appropriate, available instrument that has established reliability and validity, enables you to answer your research questions, and has been used with a population similar to yours, it is usually best to use it. Although this chapter will focus on data collected using online questionnaires, most of the information applies to data collected using other methods of collecting quantitative data.

  Pilot test and refine data collection instruments. It is always desirable to try out your data collection process and tools(s) with, at the very least, a few colleagues or friends (or their children if the study focuses on children). When possible, you also should conduct a **pilot study** with a sample similar to the one you plan to use for your study. This is especially important if you developed the procedure or instrument or if it is going to be used with a population different from the one(s) for which it was developed and on which it was previously used. Pilot study participants can provide insights into whether instructions and questions are clear, procedures are similar to those they have experienced, etc. Then, you can use the feedback to make modifications before beginning the actual data collection. Typically, the pilot data should not be added to the data collected for the study.

Download Data Collected Online

If you collected data using an online survey generator (e.g., Qualtrics), usually it is possible to download a spreadsheet with all participants' individual answers on them. You always should download a spreadsheet with **each individual's responses to each question**, rather than a summary of results and charts/figures, so that you can do inferential statistical analyses on the data. You can either download it as an SPSS file (if this option is available) or download it as an Excel

DOI: 10.4324/9781003355908-2

file and then import the file into SPSS. We will present how to download data into SPSS and then how to download to Excel and then import into SPSS.

Downloading Data Into SPSS

If possible, download your file as an SPSS data (.sav) file and save it to your computer using a file name that clearly describes the nature of the data (e.g., hsbdata for High School and Beyond data). If it is not possible to download it as an SPSS file, you can download it as an Excel work-book or Excel csv file and import it into SPSS.

When data are collected online, there usually is a column with a timestamp, and sometimes other columns with data that are not participant responses to your survey items. As long as you have participant numbers, you can delete such columns before using the data, but you usually should save this edited file with a new name, just in case you might need the information in those columns. Keeping these data may be helpful if you decide to delete participants or merge other files as these data will help with identifying the individual participants in the file if there are errors with participant numbers, demographic data, etc.

Make sure that the column headers, which are the variable names, are *short* names that *uniquely describe one and only one variable* (usually 8 letters or less is best), and the data for each partici-pant are listed as a row of data, with the same variables in the same columns for all participants. If your file does not have column headers, insert a row and add a name (no more than eight characters) in the top row for each variable. If the column headers already exist, but are longer than eight letters/characters, shorten them to names with at most eight characters. Make sure that each variable name is unique (different from all other column headers), does not have any spaces/blanks, and is a name that will help you remember the variable in the column (e.g., "Modality" for course modality; "mathgrade" for mathematics grades). Add a participant number variable and number participants using a system that ensures each participant has a unique number. Save the file with any additions and/or corrections.

Downloading Data into Excel and Then Importing Into SPSS

If the file is an Excel file (Workbook or csv), you will need to import it to SPSS:

- Open a blank SPSS data file by double clicking on IBM SPSS Statistics.
- Click on **File** in the upper left-hand corner.
- Click on **Import data → Excel (or CSV data) and select (double click on) the correct dataset**.
- Make sure your file has the correct names of the variables as the Variable Names at the tops of the columns in the Data View. Remember, these should be relatively short; if they are too long, SPSS will not be able to use them.
- Make sure your file has participants as rows and variables as columns.
- Make sure that "Read variable names from first row of data" is checked in the SPSS dialog box.
- If you want to do so, remove leading spaces and/or trailing spaces from "string" values (words for levels of nominal variables, rather than numbers).
- Make sure that, for CSV files, "Delimiter between values," "Comma" is showing.
- The other defaults are usually fine but change them if decimals are not represented as periods and double quotes are not used to delimit words that represent levels ("string variables").
- Click on **OK**.

Now that you have the file open in SPSS, make sure it is open in **Data View**. If it is in the **Variable View**, click on the Data View tab (see Chapter 1 for how to do this).

- Next, look at the data to see if you have words instead of numbers for any of your scores/ survey responses. If you do, you will want to recode them to make them numbers. One way of finding variables like that is to look at the variable view and see if the "Type" column lists any of your variables as "String" variables (variables using words as the values/responses, rather than numbers), but be careful. Do NOT just change the "Type" column entry for that variable to "numerical" without changing the actual entries in the spreadsheet. This will make your spreadsheet entries for that variable all disappear! If you have variables that use words instead of numbers to indicate the ratings or levels of the response to the item/ question on the survey, first make sure that the words referring to the values are not very long (8 characters or less is best). If they are too long, SPSS will not be able to use them. You will need to use the SPSS "Transform" program to change the words to numbers. The **Data Coding and Descriptive Analysis** section of this chapter gives suggestions for these numbers.

To transform words to numbers:

- At the top of the SPSS screen, select **Transform → Recode into Different Variables**.
- Move all of the items that have exactly the same rating scale into the box labeled **Input**

  Variable → Output Variable. Once you do this, the label on the box should change to **String Variable → Output Variable**.

- Type the names you want to call each of the new variables in the Output Variable box, one at a time. For example, if the variable with words as values is called "extra1" (for item 1 on the extraversion scale), you might call the new variable "extra1#" (for extraversion item 1 numeric). Each variable name should have no more than eight characters and cannot have any spaces in it. Be sure to also type a name that fully identifies the variable in the **Label** box to make sure you remember what that variable is. The label can have many letters, and it can have spaces.
- Click on **Change**.
- Click on **Old and New Values**.
- Put the word(s) that refer(s) to one level/value of the item into the **Old Value** box and put the number you want to refer to that level in the **New Value** box.
- Click on **Add**.
- Repeat for all of the other values for those variables/items.
- Click **Continue**.
- Click **Paste**, so that you will be able to save the syntax (the instructions to SPSS about what numbers refer to what levels/values).
- You will see a Syntax file pop up (Syntax1, if it is the first recode you are doing).
- Click on **File → Save As** and save the syntax file with a name that will tell you what variables you were recoding.
- Select the whole set of commands referring to that recode and click on the green "play button" (forward pointing triangle) to run the recode.
- Look at the **Data View** of the dataset. You should have new variable(s) at the end of the row with the new name(s) and values in it.

- Click on the **Variable View** tab at the bottom left of your screen. This will bring up a screen similar to Figure 2.1. (Or *double* click on **var** above the blank column to the far left side of the **Data View**.)

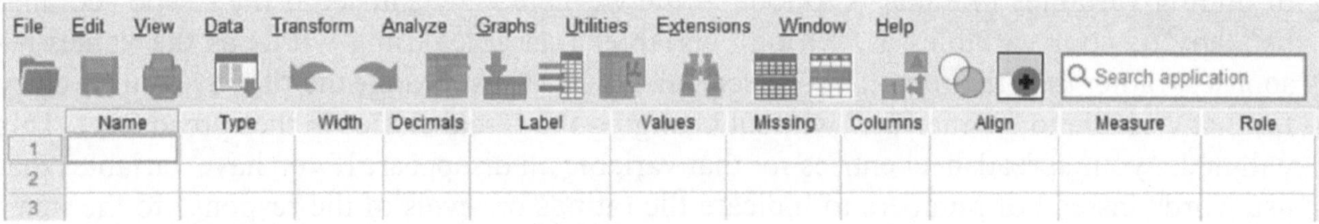

Figure 2.1 Blank variable view screen in the data editor.

- In this window, you will see 11 columns that will allow you to input the variable **Name**, **type** of variable, **width**, number of **decimals**, variable **label**, **value** labels, **missing** values other than blanks, **columns**, **align** data left or right, **measurement** type, and variable **role**.
- To see how variables are defined and labeled, it may be useful to look at some that are already labeled. We will do this by retrieving a dataset that already has variable information in it.
- Retrieve **hsbdata.sav**. It is desirable to save a copy of this file. See Chapter 1 for instructions if you need help with this or getting started. Chapter 1 also shows how to set your computer to print the syntax.
- Click on the **Variable View** tab at the bottom left of your screen. Notice the names of variables listed in the column titled **Name.** If you were going to enter data by hand into SPSS, you would start by defining the variables. Click in the first blank box in the column under **Name** in Figure 2.4. This is where you would give your variable a name. You can enter a name (8 characters or less) for your variable here to practice doing this.
- Press Enter. This will insert the program's default values for variables. You need to check to be sure these are correct for each of your variables and make changes if needed. Note that the **Type** is numeric, **Width** = 8, **Decimals** = 2, **Label** = (blank), **Values** = None, **Missing** = None, **Columns** = 8, **Align** = right, **Measure** = scale, **Role** = input.

For this assignment, we will keep the default values for **Type**, **Width**, **Columns**, and **Align**. On the **Variable View** screen, you will notice that the default for **Type** is **Numeric**. This refers to the type of variable you are entering. Usually, you will only use the **Numeric** option. Numeric means the data are numbers. **String** would be used if you input words or letters such as "M" for males and "F" for females. However, it is best not to enter words or letters because you wouldn't be able to do many statistics without recoding them as numbers. In this book, we will always keep the Type as Numeric.

We recommend keeping the **Width** at eight, and keeping the **Columns** at eight. We will always **Align** the numbers to the right. Sometimes, we will change the settings for the other columns.

- Check the **Variable View** to make sure that the level of measurement (**Measure**) is **Scale** (many values from low to high, probably normally distributed) or **Ordinal** (a few ordered values but not likely to be normally distributed), as appropriate for your data.

- It is important to add value labels for the different levels, indicating what they refer to in the **Values** column. Since this is just a made-up variable, you can make these whatever you'd like to make them.

Adding labels:

- Click on the word "None" and you will see a small blue box with three dots.
- Click on the three dots. You will then see a screen like Figure 2.2. For Likert scales, it is usually sufficient to add **value labels** for the lower and upper end of the Likert scale to help you interpret the data.

Figure 2.2 Value labels.

- Type 1 in the **Value** box in Figure 2.3.

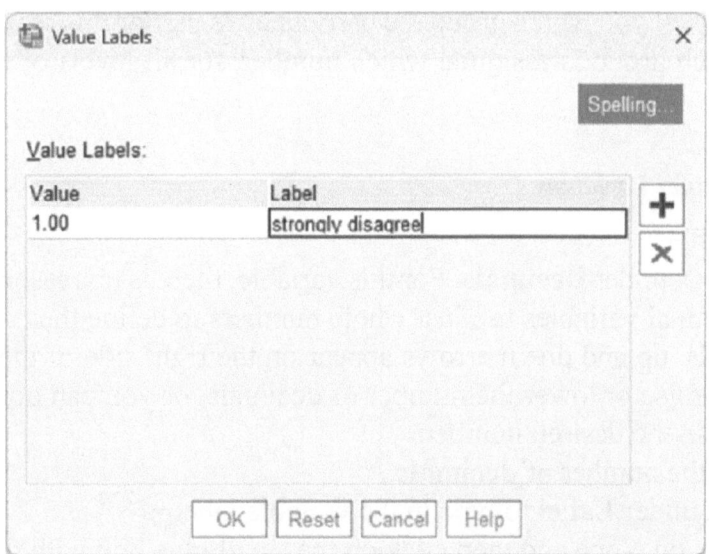

Figure 2.3 Value labels with a value.

- Type the lowest value of your Likert scale, such as *strongly disagree,* in the **Value Label** box. Press **Add**.
- Type the highest number of your Likert scale in the **Value** box and then type the highest value of your Likert scale, such as *strongly agree* in the **Value Labels** box. Your window should look like Figure 2.4 just before you click on **Add** for the second time.

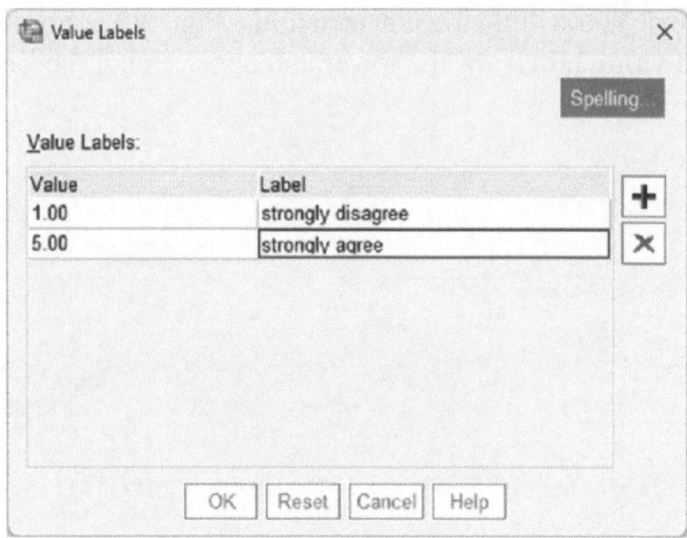

Figure 2.4 Value labels cont.

- Click on **Add**.
- Then click **OK**.

For nominal data, the numbers you assign to each group are arbitrary, yet they are of critical importance because they tell you which group the participant is in. For this reason, it is crucial to use labels to tell you which group is assigned which number, for all levels of the nominal variable. Here are the steps:

Define and Label Nominal Variables

- Highlight the variable.
- Click on the third box under **Decimals**. For this variable, there is no reason to have any decimal places because nominal variables use just whole numbers to define them. When you select the box under **Decimals**, up and down arrows appear on the right side of the box. You can either click the arrows to raise or lower the number of decimals, or you can double click on the box and manually type in the desired number.
- Select or type 0 as the number of **decimals**.
- Next, click the box under **Label** to type in the variable label.
- Under **Values**, click on None and then click on the small blue box with three dots.
- In the **Value Labels** window, type 1 in the **Value** box, type the label for the group represented by 1 in the **Value Label** box.

• Then click **Add**. Do the same for all other values of the variable. The **Value Labels** window should resemble Figure 2.6 just before you click **Add** for the last time. Save your dataset with a new name indicating that it is the updated or final version of the data.

Data Coding and Descriptive Analysis

Before computing any inferential statistics, it is necessary to code the data, enter the data into SPSS (if it is not already in an SPSS file), and then conduct descriptive data analysis as outlined in the next section. This will help you understand your data, see if there are any errors, and find out if your data meet basic assumptions for the inferential statistics that you will compute.

RULES OF THUMB FOR DATA CODING

Coding is the process of assigning numbers or symbols to the values or levels of each variable. In the previous section, you coded any "string" variables to change words to numbers. We want to present some broad suggestions or rules to keep in mind as you code all of your data. These suggestions are adapted from rules proposed in Newton and Rudestam's (2013) useful book entitled *Your Statistical Consultant*. It is important to note that the recommendations we make are those we find to be most useful in most contexts, but some researchers might propose alternatives, especially for "rules" 1, 2, 4, 5, and 7. If, after reading these guidelines, you want to code your responses differently than they appear in your file, you can do this using **TransformRecode into Different Variables**, as shown earlier in this chapter.

1. **All data should be numeric**. Even though it is possible to use letters or words (string variables) as data, it is not desirable to do so. For example, we could code *academic track* as F for fast track and R for regular track, but in order to do most statistics with SPSS, you would have to convert the letters or words to numbers. It is easier to do this conversion before analyzing the data, when you first upload or enter the data into the computer. We decided to code regular track as 1 and fast track as 0. This is called **dummy coding**. In essence, the 0 means "not regular track." Dummy coding is useful if you will want to use the data in some types of analyses and for obtaining descriptive statistics. For example, the mean of data coded this way will tell you the percentage of participants who fall in the category coded as "1," which, in this example is the percentage of participants in the regular academic track. We could, of course, code fast track as 1 and regular track as 0 and do the same thing, in which case the mean would be the percentage of students in the fast track. We could even code one academic track as 1 and the other as 2, in which case if you calculated the mean, the numbers after the decimal point would tell you the percent of participants in the group coded as "2." However, it is crucial that you be consistent in your coding and have a way to remind yourself and others of how you did the coding.
2. **Each variable for each case or participant must occupy the same column in the SPSS Data Editor**. For most SPSS procedures (**Linear Mixed Models** is an exception), it is important that data from each participant occupy only one line (row), with each column specifying one variable; moreover, for all procedures, each column must contain data on the same variable for all the participants. The SPSS data editor facilitates this by putting the variable names that you choose at the top of each column, as you saw in Figure 2.5. <u>If a variable is measured</u>

more than once for each participant (e.g., at pretest and posttest), it usually needs to be entered on the same row in separate columns with somewhat different names like *mathpre* and *math-post* or *mosaic1* and *mosaic2*. An exception to this is for the **Linear Mixed Models** program when you are doing a repeated-measures model (see Chapter 11).

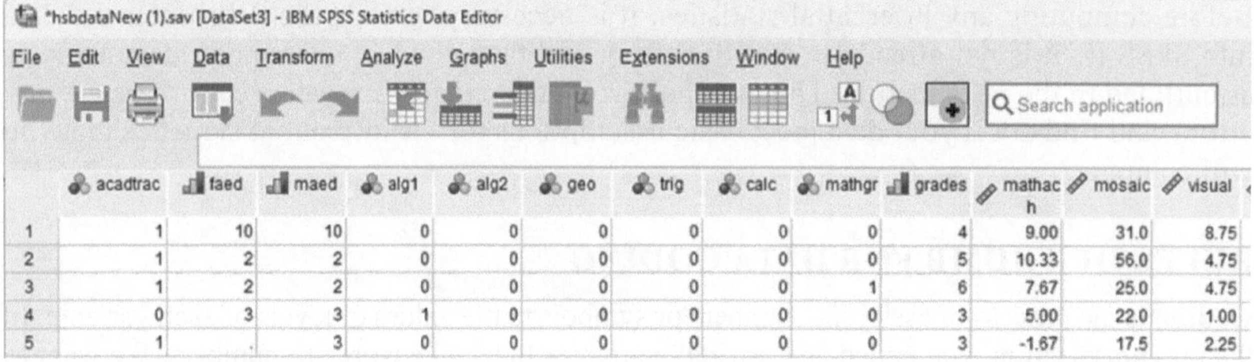

	acadtrac	faed	maed	alg1	alg2	geo	trig	calc	mathgr	grades	mathach	mosaic	visual
1	1	10	10	0	0	0	0	0	0	4	9.00	31.0	8.75
2	1	2	2	0	0	0	0	0	0	5	10.33	56.0	4.75
3	1	2	2	0	0	0	0	0	1	6	7.67	25.0	4.75
4	0	3	3	1	0	0	0	0	0	3	5.00	22.0	1.00
5	1	.	3	0	0	0	0	0	0	3	-1.67	17.5	2.25

Figure 2.5 Data editor.

3. **All values (codes) for a variable must be mutually exclusive**. That is, only one value or number can be recorded for each variable for a particular participant. Some items may allow participants to check more than one response or observers to note more than one response. In that case, the item should be divided into *a separate variable for each possible response choice*, with one value of each variable (usually coded "1") corresponding to yes (checked) and the other (usually "0") to no (not checked).

Usually, items should be phrased so that persons would logically choose only one of the provided options, and all possible options are provided. A final category labeled "other" may be provided in cases where all possible options cannot be listed, but these "other" responses are usually quite diverse and, thus, are usually not very useful for statistical purposes.

4. **Each variable should be coded to obtain maximum information**. Do not collapse categories or values when you set up the codes for them. If needed, let the computer do it later. In general, it is desirable to code and enter data in as detailed a form as available. Thus, enter item raw scores, ages (or date of birth and testing date, from which SPSS can calculate age), GPAs, etc. for each participant. It is good to ask participants to provide information that is quite specific or have observers code data in ways that are as objective as possible.

However, you should be careful not to ask questions that are so specific that the respondent may not know the answer or may not feel comfortable providing it. For example, you will obtain more specific information by asking participants to state their GPA to two decimals than if you asked them to select from a few broad categories (e.g., less than 2.0, 2.0–2.49, 2.50–2.99). However, if students don't know their exact GPA or don't want to reveal it precisely, they may leave the question blank, guess, or write in a difficult to interpret answer. These issues might lead you to provide categories that are as specific as people are likely to be able to provide (such as ranges of .25 points). Similarly, if, even after training, observers cannot reliably distinguish among the specific categories you provide (e.g., extremely happy versus very happy), the categories should be modified. Then, enter the data exactly as collected. If you find out later that some categories were not used or were used too infrequently, you can always combine several categories using an SPSS function, **Transform →Recode into Different Variables**, as shown earlier in this chapter. If you collapse categories before you enter the data, the additional information will no longer be available.

5. **For each participant, there must be a code or value for each variable, except if data are missing**. These codes should be numbers, except when the data are missing. We recommend using blanks when data are missing or unusable, because SPSS is designed to handle blanks as missing values. However, sometimes you may have more than one type of missing data, such as items left blank *and* those that had an answer that was not appropriate or usable. In this case, you may assign numeric codes such as −98 and −99 to them, but you must tell SPSS that these codes are for missing values, or SPSS will treat them as actual data. The reason it can be useful to code missing values as large negative numbers is that then it will usually be obvious that they are not being treated as missing values if your minimum, mean, etc. are very different than you would expect, given the actual values of the variable.

6. **Apply coding rules consistently for all participants**. This means that if you decide to treat a certain type of response as, say, missing for one person, you must do the same for all other participants.

7. **Use high numbers (values or codes) for the "agree more" or "more" end of a variable that is ordered. If the variable has a negative-sounding label (e.g., aggressiveness) then, higher numbers should refer to more of the trait, which would end up yielding scores in which higher = more aggressive or negative**. Sometimes you will see questionnaires that use 1 for "strongly agree" and 5 for "strongly disagree." This is not wrong as long as it is clear and consistent. However, you are less likely to get confused when interpreting your results if high values mean "more" (such as more agreement or more of the characteristic) or indicate that something was done (e.g., an algebra 1 course *was* taken).

8. **Make a document describing decision rules or codebook**. You need to make some decisions about how to code the data, especially data that are not already in numerical form. When the responses provided by participants are numbers, it makes most sense for the value for that variable in the SPSS file to be the number that was provided on the questionnaire. In such cases, the variable is said to be "self-coding." On the other hand, variables such as *academic track* or *ethnicity* have no intrinsic values associated with them, so a number has to be assigned to each level or value (see rule 1). Be sure to create a file with codes and decision rules, so that coding is consistent across all participants.

9. **Fix problems with the data**. Examine the data for incomplete, unclear, or inappropriate answers. The researcher needs to use systematic rules to handle these problems and note the decision on the master "coding instructions" file so that the same rules are used for all cases. For each type of incomplete, blank, unclear, or double answer, you need to make a rule for what to do. As much as possible, you should make these rules before data collection, but there may well be some unanticipated issues. It is important that you apply the rules consistently for all similar problems so as not to bias your results.

Missing data create problems in later data analysis and may distort the sample's representation of the population. Thus, we want to use as much of the data provided as is reasonable. The important thing, again, is that you *must* treat all similar problems the same way. If a participant answered only some of the questions or missed some procedures or a whole data collection wave, there will be lots of missing data for that person. We could have a rule such as "if half the questionnaire items are blank or invalid, we will throw out that whole questionnaire as invalid." In your research report, you should state how many questionnaires were thrown out and for what reason(s). If a participant enters a value that is not an actual option on the questionnaire for that item, but it is for an item that appears just after that item on the questionnaire, you might want to check to see if the participant missed an item and then all further items were "off" by one item. The Intermediate text (Barrett et al., 2025) provides statistical methods to impute values for missing data, which might be appropriate for participants who missed all or part of a wave of data.

10. **Clean up data**. Once you have made your rules and decided how to handle each problem, you need to go back and check and fix any data that might be incorrect, using the master coding decision rules file. Once you are comfortable with your decision rules and have corrected the data, you will do descriptive data analyses to further check for errors and to better understand your data.

For the remainder of the chapter, we will do exercises that require data, so we will use the **hsbdata** file for the problems in this chapter, but you could use the same commands with your own data. For the remaining problems in this chapter and those throughout this book, different types of analyses and plots will be generated depending on what level of measurement you have. Therefore, it is important to identify whether each of your variables is **nominal, dichotomous, ordinal**, or **normal** (SPSS uses the term **scale**; see Chapter 1 for more information about level of measurement). Keep in mind that there are times when whether you call a variable ordinal or scale might change based on your descriptive analyses. For example, a variable that you considered to be ordinal may be normally distributed and, thus, better labeled as scale. Remember that making the appropriate choice indicates that you understand your data and should help guide your selection of a statistic.

You will use three useful data transformation techniques in this chapter to get the data into the form needed to answer the research questions: **Count**, **Recode**, and **Compute**. From these operations we will produce new variables. You already have used **Transform→Recode** to convert "string" data into numeric data. You can use this same command to combine categories if you find certain categories are never or very seldom used. You can also use **Transform→Count** to create a variable that indicates how many "yes" answers someone gave. We will do this now, with the **hsbdata.sav** file. Sometimes you want to know how many items the participants have bought, how many stressors they experienced, etc. Often, the participant is asked to "check all that apply." For example, if you allow people to choose more than one ethnicity in your survey, you could use this same approach to enable you to report how many people reported more than 1 ethnicity (were multi-ethnic). In this problem, we will count the number of different math courses coded as 1, which means "taken."

Problem 2.1: Count Math Courses Taken

2.1. How many math courses (*algebra 1*, *algebra 2*, *geometry*, *trigonometry*, and *calculus*) did each of the 75 participants take in high school? **Label** your new variable.

Let's count the number of math courses (*mathcrs*) that each of the 75 participants took in high school. Select **Transform → Count Values within Cases . . .** You will see a window like Figure 2.5.

- Now, type *mathcrs* in the **Target Variable** box. This is the name for your new variable.
- Next, type *math courses taken* in the **Target Label** box. This is the label for the new variable. Then, while holding down the shift key, highlight *algebra 1*, *algebra 2*, *geometry*, *trigonometry*, and *calculus* and click on the **arrow** button to move them over to the **Numeric Variables** box. Your **Count** window should look like Figure 2.6.

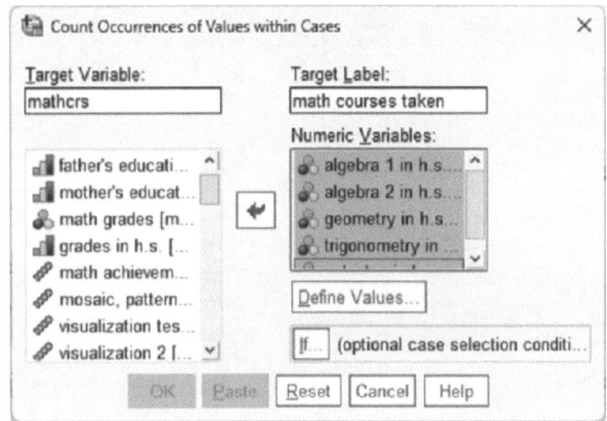

Figure 2.6 Count occurrences of values within cases.

- Click on **Define Values**.
- Type **1** in the **Value** box and click on **Add**. This sets up the computer to count how many 1s (or courses taken) each participant had. The window will now look like Figure 2.7.

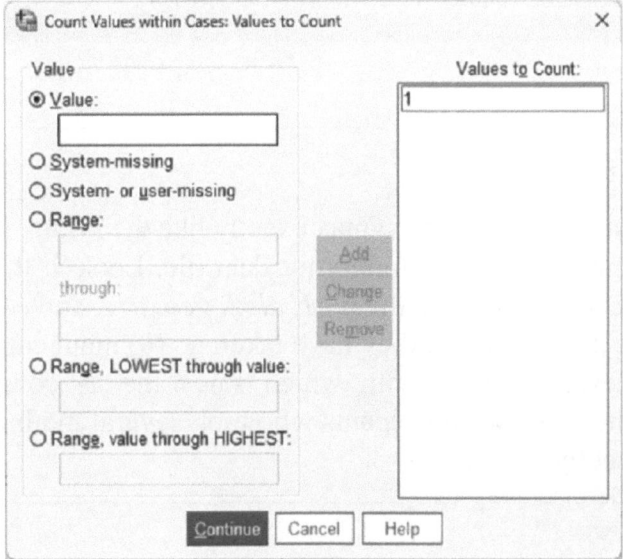

Figure 2.7 Count values within cases: values to count.

- Now click on **Continue** to return to the dialog box in Figure 2.6.
- Click on **OK**. Scores for the first ten participants on your new variable, under *mathcrs*, should look like Figure 2.8. *It is the last variable, way over to the right side of your* **Data View**, and the last variable way on the bottom of your **Variable View**.

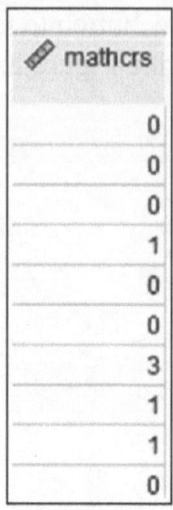

Figure 2.8 Data column.

Your output should look like the syntax in Output 2.1. There will not be any tables or plots.

Output 2.1: Counting Math Courses Taken

```
COUNT mathcrs=alg1 alg2 geo trig calc(1).
VARIABLE LABELS mathcrs 'math courses taken'.
EXECUTE.
```

Interpretation of Output 2.1

Check your syntax and counts. Is the syntax exactly like the syntax in Output 2.1? Another way to check your count is by examining your data file. Look at the first participant (top row) and notice that there are zeros in the *alg1*, *alg2*, *geo*, *trig*, and *calc* columns. The same is true for Participants 2 and 3. Thus, they have taken no (0) math courses. They should and do have zeros in the new *mathcrs* column, which is now the last column on the right. Also, it would be good to check a few participants who took several math courses just to be sure the count worked correctly.

Notice that there are no tables or figures for this output, just syntax. If you did not get any output, <u>check to make sure that you set your computer to obtain a listing of the syntax (see Chapter 1)</u>.

Problem 2.2: Recode and Relabel Mother's and Father's Education

Now we will **Recode** *mother's education* and *father's education* so that those with no postsecondary education (2s and 3s) have a value of **1**, those with some postsecondary will have **2**, and those with a bachelor's degree or more will have a value of **3**. **Recode** is used whenever there is a need to change a value or score on a variable to something different. It can be used to reverse scores (e.g., "1" becomes "7," "2" becomes "6" and so on) when a questionnaire item is "worded backwards" so that higher numbers indicate less of the variable, or when you want to change "string variable" words to numerical values, etc.

It is usually *not* desirable to dichotomize (divide into two categories) or trichotomize (divide into three categories) an ordinal or, especially, a normal/scale variable in which all of the levels are ordered correctly and are meaningfully different from one another. However, we need an independent variable with a few levels or categories to demonstrate certain analyses later, and these variables seem to have a logical problem with the ordering of the categories/values. The problem can be seen in the codebook. A value of 5 is given for two years of vocational/community college (and presumably an associate's degree), but a 6 is given for less than two years of (a four-year) college. Thus, we could have a case where a parent who went to a four-year college for a short time would be rated as having more education than a parent with an associate's degree. This would make the variable not fully ordered.

Recodes also are used to combine two or more small groups or categories of a variable so that group size will be large enough to perform statistical analyses. For example, we have only a few fathers or mothers who have a master's or doctorate so we will combine them with bachelor's degrees and call them "B.S. or more."

2.2. **Recode** *mother's* and *father's education* so that those with no postsecondary education have a value of 1, those with some postsecondary education have a value of 2, and those with a bachelor's degree or more have a value of 3. **Label** the new variables and values. Also print the frequency distributions for *maed* and *faed*.

Follow these steps:

- Click on **Transform → Recode into Different Variables** and you should get a window similar to Figure 2.9, but without the variables in the window.
- Now click on *mother's education* and then the **arrow** button.
- Click on *father's education* and the **arrow**. This will move both variables to the **Numeric Variables → Output Variable** box.
- Now highlight *faed* in the **Numeric Variable** box so that it changes color.
- Click on the **Name** box under **Output Variable**; and type *faedRevis*.
- Click on the **Label** box and type *father's educ revised*.
- Click on **Change**. Did you get *faed → faedRevis* in the **Numeric Variable → Output Variable** box as in Figure 2.9?

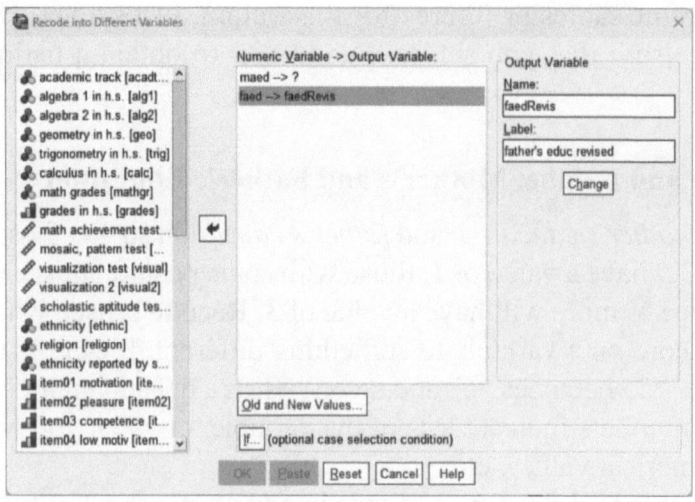

Figure 2.9 Recode into different variables.

Now repeat these procedures with *maed* in the **Numeric Variable → Output Variable** box.

- Highlight *maed*.
- Click on **Name** under **Output Variable**; type *maedRevis*.
- Click **Label**; type *mother's educ revised*.
- Click **Change**.
- Then click on **Old and New Values** to get a window similar to Figure 2.10, but without the values in it.

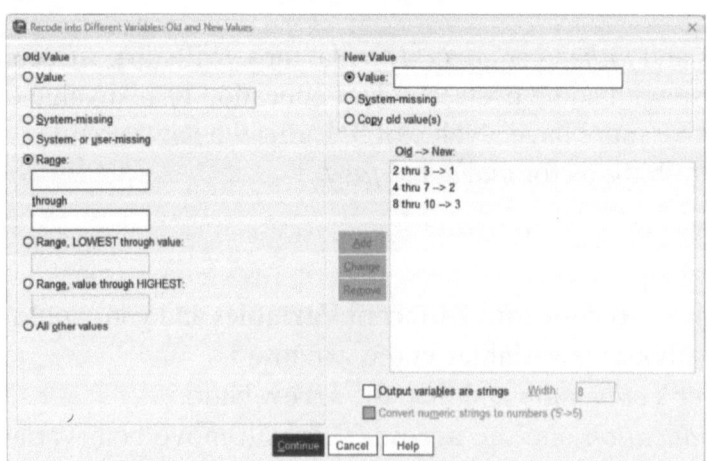

Figure 2.10 Recode.

- Click on **Range** and type **2** in the first box and **3** in the second box.
- Click on **Value** (under **New Value** on the right) and type **1**.
- Then click on **Add**.
- Repeat these steps to change old values **4** through **7** to a new **Value** of **2**.

- Then **Range: 8** through **10** to **Value: 3**. Does it look like Figure 2.10?
- If it does, click on **Continue**.
- Finally, click on **OK**.

Check your **Data View** to see if *faedRevis* and *maedRevis*, with numbers ranging from 1 to 3, have been added on the far right side. These are **ordinal** variables and will be used as **input** or independent variables so you should adjust the **Variable View** accordingly.

Next, check the data file for a few participants to see if the recodes were done correctly. For example, the first participant had 10 for *faed*, which should be 3 for *faedRevis*. Is it? Check a few more to be sure or compare your syntax file with the one in Output 2.3.

- Now, we will **label** the new (1, 2, 3) values.
- Go to your hsbdata file and click on **Variable View** (it is in the bottom left corner).
- In the *faedRevis* variable row, click on **None** under the **Values** column and then ⬚ to get a window similar to Figure 2.11.
- Click on the **Value** box and type **1**.
- Type *HS grad or less* where it says **Label**.
- Click on **Add**.
- Then click on the **Value** box again and type **2**.
- Click on the **Label** box and type *Some College*.
- Click on **Add**.
- Click once more on the **Value** box and type **3**.
- Click on the **Label** box and type *BS or More*.
- Again, click on **Add**. Does your window look like Figure 2.11? If so,
- Click on **OK**.

Important: You have only labeled *faedRevis* (*father's educ revised*). You need to repeat these steps for *maedRevis*. Do **Value Labels** for *maedRevis* on your own.

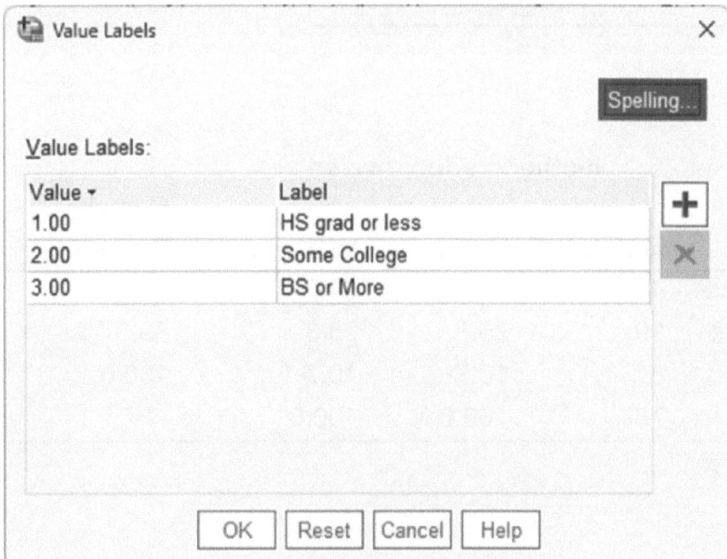

Figure 2.11 Value labels.

Now that you have recoded and labeled both *faedRevis* and *maedRevis*, you are ready to run a frequency distribution for them. Use the following commands:

- Select **Analyze → Descriptive Statistics → Frequencies**.
- Now *highlight faedRevis* and *maedRevis* (at the bottom of the list) in the left box by highlighting one, holding down the Ctrl key, and highlighting the other.
- Click on the **arrow** button pointing right.
- Be sure the **Display frequency tables box** is checked.
- Do not click on **Statistics** because we do not want to select any this time, as we will do descriptive statistics later. You do want to check descriptive statistics on new variables, but we will do this using a different program. But you could click on **Statistics** if you wanted to quickly find out descriptive statistics like **mean, SD, and skewness**.
- Click on **OK**.

Output 2.2: Recoding Mother's and Father's Education and Computing Frequencies

```
RECODE maed faed (2 thru 3=1) (4 thru 7=2) (8 thru 10=3) INTO
maedRevis faedRevis.
VARIABLE LABELS maedRevis "mother's educ revised" /faedRevis
"father's educ revised".
EXECUTE.
FREQUENCIES VARIABLES=maedRevis faedRevis
  /ORDER=ANALYSIS.
```

Frequencies

Statistics

		mother's educ revised	father's educ revised
N	Valid	75	73
	Missing	0	2

Frequency Table

mother's educ revised

		Frequency	Percent	Valid Percent	Cumulative Percent
Valid	HS grad or less	48	64.0	64.0	64.0
	Some College	19	25.3	25.3	89.3
	BS or More	8	10.7	10.7	100.0
	Total	75	100.0	100.0	

father's educ revised

		Frequency	Percent	Valid Percent	Cumulative Percent
Valid	HS grad or less	38	50.7	52.1	52.1
	Some College	16	21.3	21.9	74.0
	BS or More	19	25.3	26.0	100.0
	Total	73	97.3	100.0	
Missing	System	2	2.7		
Total		75	100.0		

Interpretation of Output 2.2

The syntax shows that you have recoded *father's* and *mother's education* so that 2 and 3 become 1, 4 through 7 become 2, and 8 through 10 become 3. The new variable names are *faedRevis* and *maedRevis*, and the labels are *father's educ revised* and *mother's educ revised*. Remember, it is crucial to check some of your recoded data to be sure that it worked the way you intended. The first **Frequencies** table shows the number of students who had valid data for each variable and the number who had missing data. Note that father's education was unknown/missing for two students. The second table shows the frequency distribution for the revised mother's education variable. Note that most (64%) of the mothers had only a high school education or less. The third table shows the distribution of father's education. Note that two are missing so the **Percent** is different from the **Valid Percent**, which is the percentage of those with valid data.

Problem 2.3: Reverse Low Pleasure Items for Pleasure Scale Score

We want to **Compute** the average "pleasure from math" scale score (*pleasure scale*) from the items that ask about pleasure: *item02*, *item06*, *item10*, and *item14* <u>after</u> reversing (**Recoding**) *item06* and *item10*, which are <u>low</u> pleasure items (see the codebook in Chapter 1). Averaging a set of items usually results in a more reliable measure than a single item would provide. We will keep both the new *item06r* and *item10r* and old (*item06* and *item10*) variables to check the recodes and to play it safe. Then we will **Label** the new computed variable as *pleasure scale*.

2.3. **Compute** the average *pleasure scale* from *item02*, *item06*, *item10*, and *item14* <u>after</u> reversing (use the **Recode** function) *item06* and *item10*. Name the new computed variable *pleasure* and label its highest and lowest values.

- Click on **Transform** → **Recode Into Different Variables**.
- Click on **Reset** to clear the window of old information as a precaution.
- Click on *item06*.
- Click on the **arrow** button.
- Click on **Name** under **Output Variable** and type *item06r*.
- Click on **Label** and type *item06 reversed*.
- Finally click on **Change**.
- Now repeat these steps for *item10*. Does it look like Figure 2.12?
- Click on **Old and New Values**.
- Now click on the **Value** box (under **Old Value**) and type **4**.
- Click on the **Value** box under **New Value** and type **1**.
- Click on **Add**.

This is the first step in recoding. You have told the computer to change values of 4 to 1. Now do these steps over to recode the values **3** to **2**, **2** to **3**, and **1** to **4**. If you did it right, the screen will look like Figure 2.13 in the **Old --> New** box. <u>Check your box carefully to be sure the recodes are exactly like</u> Figure 2.13.

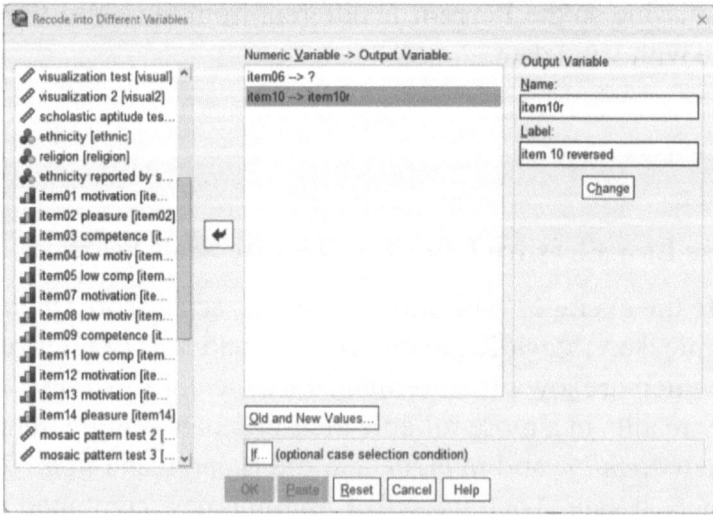

Figure 2.12 Recode into different variables.

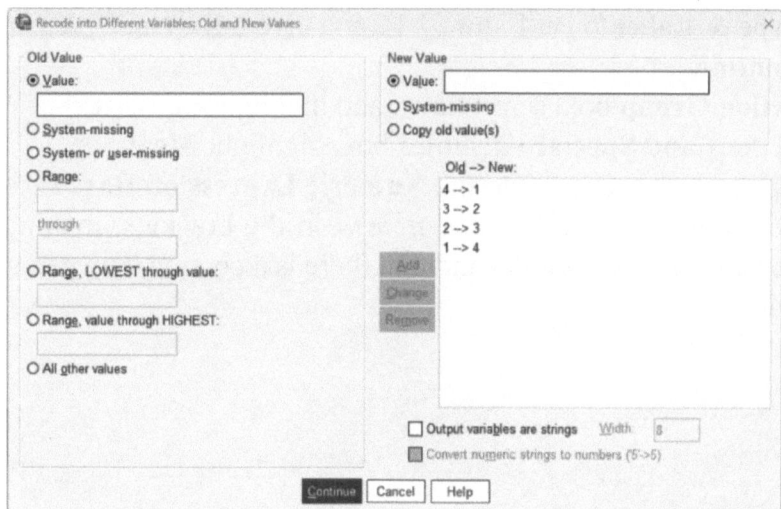

Figure 2.13 Recode: Old and new values.

- Click on **Continue** and then **OK**.

Now check your **Data** file to see if there is an *item06r* and an *item10r* in the last two columns with numbers ranging from 1 to 4. To double check the recodes, compare the *item06* and *item10* columns in your data file with the *item06r* and *item10r* columns for a few participants. Also, you should check your syntax file with Output 2.4.

Output 2.3: Recoding to Reverse Pleasure Scale Items

```
RECODE item06 item10 (4=1) (3=2) (2=3) (1=4) INTO item06r item10r.
VARIABLE LABELS item06r 'item06 reversed' /item10r 'item10 reversed'.
EXECUTE.
```

Now let's compute the average *pleasure scale*.

Problem 2.4: Compute Pleasure Scale With the Mean Function

We have decided to **Compute** the average *pleasure scale* score using the **Mean** function. The **Mean** function is a quick way to average item scores to create an overall score. If a participant is missing one or more items, it will be computed using whichever items that participant has. This reduces missing data, but it could introduce bias if certain items are more likely to be left blank.

2.4. **Compute** *pleasure* using the **Mean** function. You could have typed in the formula for mean in the **Numeric Expression** box instead of using the **Mean** function, but then if any *pleasure* item were missing, that participant would have no *pleasure* score.

- Click on **Transform → Compute Variable** to get Figure 2.14.
- In the **Target Variable:** box, type *pleasure*.

- Click on **Type & Label** to get Figure 2.15 and give it the name *pleasure scale*.
- Click on **Continue**.
- In the **Function Group** box, scroll down and highlight **Statistical**.
- In the **Functions and Special Variables** box, highlight **Mean**.
- Click the up arrow to move it into the **Numeric Expression Box**.
- Enter *item02*, *item06r*, *item10r*, and *item14* in the brackets, making sure you use the reversed items for items 6 and 10 and that there is a <u>comma between each of the items</u>.
- Click on **OK**.

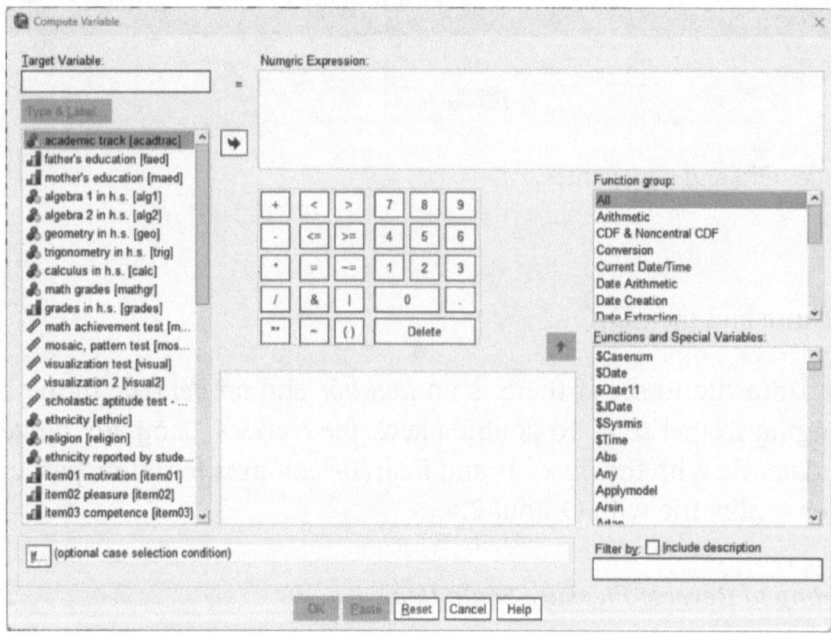

Figure 2.14 Compute new variable.

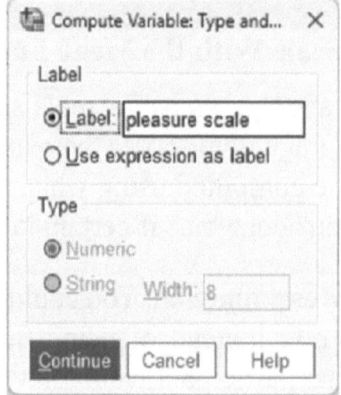

Figure 2.15 Label new variable.

Also, you should label, at least, the highest and lowest **Values**.

- Go to the **Variable View** (in the bottom left-hand corner) of your hsbdata file.
- In the *pleasure scale* variable row, click on **None** under the **Values** column, and then the three dots [...]
- Type 1, then *very low*, and click **Add**.
- Type 4, then *very high*, and click **Add**. Click on **OK**. (Note you can get values of 2.5 and 3.5, etc., so leave the decimals at 2.)

Output 2.4: Computation of Pleasure Scale

```
COMPUTE pleasure=MEAN(item02,item06r,item10r,
item14).
VARIABLE LABELS pleasure 'pleasure scale'.
EXECUTE.
```

> The Mean function computes an average score for each participant who has a score for any of the variables used.

Interpretation of Output 2.4

Note that you have created another new variable (*pleasure*), which is the last one you will create for the hsbdata. It can be found as the last variable in the **Variable View** and in the far right column of the **Data View**.

You will see when we print the **Descriptives** in Problem 2.5 that all 75 participants have a *pleasure scale* value because none of them are missing all pleasure scale items. It seems reasonable to use *pleasure scale* even if participants are missing an item or two. Because we assume that all of the *pleasure scale* items measure the same thing, by using the **Mean** function, we can increase the likelihood that we have a *pleasure scale* score for almost all participants. On the other hand, the **Mean** function should be used cautiously because, for example, if a student answered only one of the four *pleasure scale* items, they would get an average *pleasure score* based on that one item. If the item was not representative (i.e., usually rated higher or lower than the others), then the *pleasure* score would be misleading. SPSS allows you to specify the minimum number of variables that must have valid (non-missing) values by typing .n after **MEAN**. For example, if you decided that at least three of the pleasure scale items must have data, you would type MEAN.3 (*item02*, *item06r*, *item10r*, *item14*).

Checking for Errors and Normality for the New Variables

Anytime that you download or enter new data or transform or create new variables from existing variables, it is crucial to check the data for errors. Moreover, it is also important to check data to make sure that they meet the assumptions of any tests you plan to run before you run the tests. We will describe here some of the most important statistical assumptions of statistical tests you may run and how to check to make sure you meet those assumptions. The same descriptive analyses will also help you check for obvious errors in your data.

Statistical Assumptions

Every statistical test has assumptions. Statistical assumptions are much like the directions for appropriate use of a product found in an owner's manual. **Assumptions** explain when it is and isn't reasonable to perform a specific statistical test. When the *t* test was developed, for example, the person who developed it needed to make certain assumptions about the distribution of scores, etc., in order to be able to calculate the statistic accurately. If these assumptions are not met, the value that SPSS calculates, which tells the researcher whether the results are statistically significant, will not be completely accurate and may even lead the researcher to draw the wrong conclusion about the results. In each chapter, appropriate statistics and their assumptions are described.

Parametric tests. These include most of the familiar statistics (e.g., *t* test, analysis of variance, Pearson correlation, and almost all of the statistics discussed in Chapters 6 and 7). They usually have more assumptions than nonparametric tests. Parametric tests were designed for data that have certain characteristics, including approximately normal distributions.

Some parametric statistics have been found to be "robust" to violations of one or more of their assumptions. **Robust** means that the assumption can be violated to some extent without damaging the validity of the statistic. For example, one assumption of ANOVA is that the dependent variable is normally distributed for each group. Statisticians who have studied these statistics have found that even when data are not completely normally distributed at the group level (e.g., they are somewhat skewed), they still can be used under many circumstances. However, recent research suggests that most educational test score data are non-normal, this non-normality can pose methodological problems for small sample sizes, and nonparametric tests may be more accurate for nonnormal data even with large sample sizes. Moreover, unfortunately, tests of normality, while more accurate than simple visual examination of graphs of the distribution, are based on rejection of a null hypothesis of normality. As a result, as for other inferential tests, they are more likely to be statistically significant for large samples, which are more robust to violations of normality, than for small samples. For these reasons, we recommend examining skewness as the most important check on normality.

Nonparametric tests. These tests (e.g., chi-square, Mann-Whitney U, Spearman rho) have fewer assumptions and often can be used when the assumptions of a parametric test are violated. For example, they do not require normal distributions of variables or homogeneity of variances. Unfortunately, nonparametric tests to use in place of intermediate statistics discussed in our companion intermediate book are not always available in SPSS. We will describe nonparametric options when available and appropriate.

Common Assumptions

Independence of observations. A critical assumption for many statistics in this book is independence of observations. For example, if you know one participant's value on a variable (e.g., *competence*), then this should not help you to guess the value of that variable for any other particular participant. Sometimes, this assumption is violated because one's procedures for sampling participants create systematic bias. For example, "snowball sampling," in which participants recommend other participants for the study, is likely to lead to non-independence of observations because participants are likely to recommend people who are similar to themselves. Obviously, members of the same family, or the same person measured on more than one occasion, do not comprise independent observations. There are particular methods (matched samples or "repeated measures" methods) designed to deal with the non-independence of family members or the same person measured several times or participants who are matched on some characteristic. However, these procedures are not appropriate for snowball sampling, in which different participants refer different numbers of other participants, who may or may not disclose who referred them. In general, snowball sampling should be avoided for studies relying on quantitative methods such as those in this book. Similarly, posting a survey on specialized websites may be appropriate if the website targets the population to whom you wish to generalize, but not if you wish to generalize to a larger or different population.

Homogeneity of variances. Both the *t* test and ANOVA may be affected if the variances (standard deviation squared) of the groups to be compared are substantially different, especially if the number of participants in each group differs markedly. Thus, this is often a critical assumption to meet or correct for. Fortunately, SPSS provides the Levene test to check this assumption, and it provides ways to adjust the results of a *t* test if the variances are significantly different. Unfortunately, it does not provide a corrected version of factorial ANOVA, although it does enable a correction for heterogeneity of variances for comparisons among means in One-Way ANOVA.

Normality. As mentioned previously, many parametric statistics assume that certain variables are distributed approximately normally. That is, the frequency distribution would look like a symmetrical bell-shaped or normal curve, with most subjects having values in the midrange and with a smaller number of subjects with high and low scores. A distribution that is asymmetrical with more high than low scores (or vice versa) is **skewed**. Thus, it is important to check the skewness value. A distribution with especially heavy tails (higher frequency in the tails (highest and/or lowest, rather than middle scores) of the distribution or with much thinner than usual tails has difficulties with **kurtosis.** If sufficiently extreme, these non-normal distributions can make the probability of Type I error higher than indicated by the *p* value from the statistical test (**Sig.** in SPSS). There can also be difficulties with Type II error when the observed distribution is very different from the normal distribution. When normality assumptions are violated, the general linear model and other commonly used tests can produce inflated Type I and Type II errors, as well as other undesirable properties (e.g., Bishara & Hittner, 2012; Kelley, 2005). Most statistics books do not provide advice about how to decide whether a variable is at least approximately normal. If the sample is sufficiently large, the Central Limit Theorem predicts that the distribution will approximate normality. However, real data often show non-normality, even with extremely large samples (Ho & Yu, 2015). SPSS provides three of the most common normality tests, Kolmogorov-Smirnov (K-S) test), Lilliefors corrected K-S test (7, 10), and Shapiro-Wilk. A problem with these methods is that with large samples, most variables would be found to be nonnormal, and with small samples, most variables would not differ significantly from normal, despite the fact that actually deviations from normality are more likely and more problematic with smaller samples. A simpler guideline is that if the skewness is less than plus or minus one ($< +/-1.0$), the variable is at least approximately normal. SPSS enables you to check both skewness and "excess" kurtosis (a measure that, like the skewness statistic, should be zero if there is a normal distribution). Although both of these involve deviations from normality, skewness seems to have a greater impact on results (e.g., Arnau et al., 2013). There are also several other ways to check for normality. In this chapter we will look at two graphical methods: boxplots and frequency polygons. As mentioned, simply examining plots is not as accurate as examining the distribution quantitatively. Although both *t* tests (if two-tailed) and ANOVA are quite robust to violations of normality, it may be preferable to use either nonparametric statistics or to transform data if nonnormality is extreme, especially with relatively small samples.

Linearity. Linearity is the assumption that two variables are related in a linear fashion. If variables are linearly related, then when plotted in a scatterplot, the data will fall in a straight line or in a cluster that is relatively straight. Sometimes, if the data are not linearly related (i.e., the plot looks curved) the data can be transformed to make the variables linearly related.

Checking for Errors and Assumptions With Ordinal and Scale Variables

The level of measurement of a variable you are exploring (whether it is nominal, ordinal, dichotomous, or normal/scale) influences the type of descriptive data analysis you will want to do. Thus, we have divided this chapter by the measurement levels of the variable because certain descriptive statistics will not make sense for some types of variables, (e.g., a mean for a polychotomous nominal variable, or a boxplot for a dichotomous variable). Remember that the researcher has labeled the type of measurement as either nominal, ordinal, or scale when completing the **SPSS Data Editor Variable View**. Remember also that <u>we decided to label dichotomous variables</u> **nominal**, and <u>variables that we assumed were normally distributed were labeled</u> **scale**.

For all of the remaining examples in this chapter, we will be using the **hsbdata.sav** file. See Chapter 1 for instructions if you need help with this or getting started with SPSS. Chapter 1 also shows how to set your computer to print the SPSS syntax on the output.

- Retrieve **hsbdata.sav** from the website if you have not already done so. It is desirable to make a working copy of this file.

Problem 2.5: Descriptive Statistics for Ordinal and Scale Variables

For the HSB variables that were labeled as ordinal or scale in the SPSS Variable View, it is important to see if the means make sense. Are they close to what you expected? Also, examine the minimum and maximum values of the data, and check the shape of the distribution (i.e., skewness value).

2.5a. Is the central tendency, variability, range of scores, and the shape of the distribution for each of the ordinal and scale variables reasonable given the possible values of those variables? Which of these variables are normally distributed?

One way to check these is with the SPSS **Explore** command. It is important to examine your data to see if the variables are approximately normally distributed, an assumption of most of the parametric inferential statistics that we will use. To understand if a variable is normally distributed, we compute the skewness statistic, which helps determine how much a variable's distribution deviates from the distribution of the normal curve. Skewness refers to the lack of symmetry in a frequency distribution. Distributions with a long "tail" to the right have a positive skew and those with a long tail on the left have a negative skew. If a frequency distribution of a variable has a large skewness absolute value (larger than 1 or less than –1), that variable is said to deviate from normality. Some of the statistics that we will use later in the book are robust or quite insensitive to violations of normality. Thus, we will assume that it is okay to use them to answer most of our research questions as long as the variables are not extremely skewed. In the case when variables are extremely skewed, you can transform the variables (see Intermediate book) to hopefully reduce the skewness and use parametric statistics. Or an appropriate nonparametric statistic could be used if you do not want to or cannot perform such a transformation or it fails to correct the skewness.

The **Explore** command enables you to calculate almost any descriptive statistic you want to do and to include it in one output. You could instead run the **Frequencies** program because you can get most of the same statistics and/or can calculate them with that program. However, the **Frequencies** program is best to use when you want to see a frequency distribution, which is usually not useful for continuous variables. We will use the Frequencies command later in the chapter for variables with a few levels, to show you how to use this program to calculate descriptive statistics; however, we recommend using Explore whenever there are four or more levels of ordered data and you either want to compare descriptive statistics on that variable for two or more levels of a nominal/grouping variable or when you think the variable could be non-normal. Now we will compute the mean, median, standard deviation, interquartile range, skewness, "excess" kurtosis, minimum, and maximum for all participants or cases on all the variables that were called ordinal or scale under measure in the **SPSS Data Editor Variable View**. We will <u>not</u> include the nominal variables (*ethnicity* and *religion*) or *algebra1, algebra2, geometry, trigonometry, calculus,* and *math grades*, which are dichotomous variables. If the variables you considered ordinal or originally thought would be "scale" (normally distributed) turn out empirically to be non-normally distributed, you will want to report median and interquartile range when you write about them.

First, we will compute descriptive statistics for some of the variables we have marked as **ordinal** variables. These include *father's education, mother's education,* and *grades in h.s.* Use these steps:

To get Output 2.5a, select:

- **Analyze → Descriptive Statistics → Explore**
- While holding down the control key (i.e., the key marked "Ctrl"), click on the following variables in the left box that we called **ordinal** so that they are highlighted. These include *father's education, mother's education,* and *grades in h.s.* This analysis will not only provide descriptive statistics for the variables, it will also help us determine if any of these

variables can be considered scale instead of ordinal by assessing normality of the variables' distribution.

- Click on the **arrow** button pointing right by the **Dependent List** box to move the variable names into the box.
- Be sure that all of the requested variables have moved into the Dependent List box.
- For this problem, leave the **Factor List** box empty. You would use this if you wanted to compare the statistics and distributions for the different groups comprising a nominal variable (e.g., if you wanted to compare different ethnicity groups on the variable).
- Click on **Statistics**. Be sure that **Descriptives** has a check next to it. Also check **Outliers**.
- Click on **Plots. Stem and Leaf** should already be checked. In addition, select **Histogram** so that each has a check.
- Click on **Continue** and then click on **OK**.
- We will now demonstrate how to get SPSS to modify the output table so that it is in a format more similar to APA format. Unfortunately, it usually will require further modifications to make it truly an APA-formatted table. You may decide it is easier to just create the APA table and type in the numbers from the output rather than getting SPSS to create the modified table and then modifying it.
- Double click on the descriptives table in the output.
- Click on **FormatTableLooks.**
- Select **APATimesRoma_12pt.**
- Click on **Reset all cell formats to the TableLook.**
- Click on **OK.**
- You should see an (almost) APA formatted table.
- Highlight the portion you want to import into your manuscript. Right-click on the highlighted table, click on Copy, then navigate to your manuscript page or to a blank document. Right-click to paste the text as it appears in the origin document. Now you can close the Chart Editor by clicking on the x in the upper right corner.
- Click out of the editing screen by clicking on the X in the upper right corner.

Compare your output with Output 2.5a.

Output 2.5a: Descriptives for the Ordinal Variables

```
EXAMINE VARIABLES=faed maed grades
  /PLOT BOXPLOT STEMLEAF HISTOGRAM
  /COMPARE GROUPS
  /STATISTICS DESCRIPTIVES EXTREME
  /CINTERVAL 95
  /MISSING LISTWISE
  /NOTOTAL.
```

Descriptives

			Statistic	Std. Error
father's education	Mean		4.73	.331
	95% Confidence Interval for Mean	Lower Bound	4.07	
		Upper Bound	5.39	
	5% Trimmed Mean		4.58	
	Median		3.00	
	Variance		8.007	
	Std. Deviation		2.830	
	Minimum		2	
	Maximum		10	
	Range		8	
	Interquartile Range		6	
	Skewness		.684	.281
	Kurtosis		−1.047	.555
mother's education	Mean		4.14	.265
	95% Confidence Interval for Mean	Lower Bound	3.61	
		Upper Bound	4.66	
	5% Trimmed Mean		3.96	
	Median		3.00	
	Variance		5.120	
	Std. Deviation		2.263	
	Minimum		2	
	Maximum		10	
	Range		8	
	Interquartile Range		3	
	Skewness		1.083	.281
	Kurtosis		.059	.555
grades in h.s.	Mean		5.70	.182
	95% Confidence Interval for Mean	Lower Bound	5.34	
		Upper Bound	6.06	
	5% Trimmed Mean		5.74	
	Median		6.00	
	Variance		2.408	
	Std. Deviation		1.552	
	Minimum		2	
	Maximum		8	
	Range		6	
	Interquartile Range		2	
	Skewness		−.326	.281
	Kurtosis		−.707	.555

Interpretation of Output 2.5a

This output provides descriptive statistics for some of the variables we originally labeled as **ordinal**. Notice that the variables are listed down the left column of the outputs and the requested descriptive statistics are listed down the next column. The descriptive statistics included in the output are mean, 95% confidence interval for mean, 5% trimmed mean, median, variance, standard deviation, minimum, maximum, range, interquartile range, skewness, and kurtosis.

Note, in the **Case Processing** table in your output (not shown here), that the **Valid N (listwise)** is 69 rather than 75, which is the number of participants in the data file. This is because the listwise *N* includes only the persons with *no* missing data on **any** variable requested in the output. Notice that several variables (e.g., *father's education*) each have one or two participants with missing data.

Using your output to check your data for errors. For these ordinal variables, check to make sure that all **Means** seem reasonable. That is, you should check your means to see if they are within the ranges you expected (given the information in your codebook and your understanding of the variable). Next, check to see if the mean and median are similar; this is your first indication that the data are reasonably normally distributed. Then, check the output to see that the **Minimum** and **Maximum** are within the appropriate range for each variable. If the minimum is smaller or the maximum is bigger than you expected (e.g., 1 or 15 for father's education, which has a range of 2–10 for possible values), then you should suspect that there was an error somewhere, and you need to check it out. You can also check the **Extreme Values** table to see if any scores are outside the values that are actually possible for each variable. Another thing to check is how many have missing data. If it happens that you have more participants missing than you expected, check the original data to see if some were entered incorrectly or inadvertently omitted. If you had calculated the scores you are looking at and you had more missing data than expected, you would also want to check to make sure the calculations were done correctly.

Using the output to check assumptions. The main assumption that we can check from this output is normality. We are only doing this for ordinal variables with at least five levels, because those with four or fewer levels would usually not be considered to be scale even if they were not very skewed. Usually, if a variable has fewer than five levels, it is best to view it as ordinal; however, it is important to realize that this is our guideline rather than a true requirement. Under some circumstances, variables that have four levels representing reasonably equidistant points along an underlying continuous distribution might be considered normal or scale if they are normally distributed, even though they are measured only on a four-point scale.

From Output 2.5a, we can see that, of these variables with five or more levels that we called **ordinal**, two of them (*father's education, grades in h.s.*) are approximately normally distributed; that is, they have five or more levels and have skewness values between −1 and 1. Thus, we can assume that they are more like scale variables, and we can use inferential statistics that have the assumption of normality. To better understand how these variables can be used in analyses, you can change the **Measure** column in the **Variable View** so that these variables are labeled as **scale**. Note that *mother's education*, with a skewness statistic of 1.12, is more skewed than is desirable but is not grossly skewed. Given that *father's education* is not skewed, you might decide to treat both as scale so you can do parametric analyses with both.

Next, we will run **Explore** for the <u>variables</u> we originally labeled as <u>scale</u>: *math achievement, mosaic, visualization, visualization retest, scholastic aptitude test-math, mosaic2, mosaic3, competence, motivation,* and *pleasure scales.* Note that these variables have the symbol ✐ next to them. It is important to test our assumption that these variables are normally distributed. In addition, since we calculated the pleasure scale from items, we will want to check the minimum and maximum to make sure that they are within the range of possible values for this variable. Since it is an average of items with values from 1 through 4, the minimum value should not be lower than 1, and the maximum value should not be higher than 4.

2.5b. What is the distribution of values for variables in our dataset that we have conceptualized as approximately normal or scale?

To get Output 2.5b, select the following:

- **Analyze → Descriptive Statistics → Explore** . . .
- Click on **Reset** to move the ordinal variables back to the left. This also deletes what we chose under **Options**.
- Highlight *math achievement, mosaic, visualization, visualization retest, scholastic aptitude test-math, mosaic2, mosaic3, competence, motivation,* and *pleasure scale* and move them to the **Dependent List** box.
- Click on **Statistics** and repeat what you did in Problem 2.5a.
- Click on **Plots** and repeat what you did for 2.5a.
- Click on **Continue** and then **OK**.

Compare your output with Output 2.5b.

Output 2.5b: Descriptives for Variables Labeled as Scale

```
EXAMINE VARIABLES=mathach mosaic visual visual2 satm
mosaic2 mosaic3 competence motivation pleasure
  /PLOT BOXPLOT STEMLEAF HISTOGRAM
  /COMPARE GROUPS
  /STATISTICS DESCRIPTIVES EXTREME
  /CINTERVAL 95
  /MISSING LISTWISE
  /NOTOTAL.
```

			Statistic	Std. Error
math achievement test	Mean		12.6479	.81161
	95% Confidence Interval for Mean	Lower Bound	11.0292	
		Upper Bound	14.2666	
	5% Trimmed Mean		12.7131	
	Median		13.0000	
	Variance		46.768	

	Std. Deviation		6.83875	
	Minimum		-1.67	
	Maximum		23.67	
	Range		25.33	
	Interquartile Range		11.67	
	Skewness		.008	.285
	Kurtosis		-1.035	.563
mosaic, pattern test	Mean		26.937	1.0928
	95% Confidence Interval for Mean	Lower Bound	24.757	
		Upper Bound	29.116	
	5% Trimmed Mean		26.700	
	Median		26.000	
	Variance		84.785	
	Std. Deviation		9.2079	
	Minimum		-4.0	
	Maximum		56.0	
	Range		60.0	
	Interquartile Range		8.0	
	Skewness		.348	.285
	Kurtosis		3.215	.563
visualization test	Mean		5.2641	.47245
	95% Confidence Interval for Mean	Lower Bound	4.3218	
		Upper Bound	6.2064	
	5% Trimmed Mean		5.0737	
	Median		4.7500	
	Variance		15.848	
	Std. Deviation		3.98096	
	Minimum		-.25	
	Maximum		14.75	
	Range		15.00	
	Interquartile Range		6.25	
	Skewness		.518	.285
	Kurtosis		-.457	.563
visualization 2	Mean		5.1831	.45527
	95% Confidence Interval for Mean	Lower Bound	4.2751	
		Upper Bound	6.0911	
	5% Trimmed Mean		5.0141	

	Median		5.0000	
	Variance		14.716	
	Std. Deviation		3.83614	
	Minimum		.00	
	Maximum		15.00	
	Range		15.00	
	Interquartile Range		6.00	
	Skewness		.484	.285
	Kurtosis		-.574	.563
scholastic aptitude test - math	Mean		490.99	11.528
	95% Confidence Interval for Mean	Lower Bound	468.00	
		Upper Bound	513.98	
	5% Trimmed Mean		490.92	
	Median		500.00	
	Variance		9434.728	
	Std. Deviation		97.133	
	Minimum		250	
	Maximum		730	
	Range		480	
	Interquartile Range		90	
	Skewness		.111	.285
	Kurtosis		.744	.563
mosaic pattern test 2	Mean		26.9718	1.05940
	95% Confidence Interval for Mean	Lower Bound	24.8589	
		Upper Bound	29.0847	
	5% Trimmed Mean		26.5970	
	Median		25.0000	
	Variance		79.685	
	Std. Deviation		8.92664	
	Minimum		.00	
	Maximum		56.00	
	Range		56.00	
	Interquartile Range		8.00	
	Skewness		.760	.285
	Kurtosis		3.305	.563
mosaic pattern test 3	Mean		26.0845	1.05897
	95% Confidence Interval for Mean	Lower Bound	23.9724	
		Upper Bound	28.1966	

	5% Trimmed Mean		25.6847	
	Median		25.0000	
	Variance		79.621	
	Std. Deviation		8.92308	
	Minimum		.00	
	Maximum		55.00	
	Range		55.00	
	Interquartile Range		10.00	
	Skewness		.799	.285
	Kurtosis		2.803	.563
competence scale	Mean		3.2887	.07987
	95% Confidence Interval for Mean	Lower Bound	3.1294	
		Upper Bound	3.4480	
	5% Trimmed Mean		3.3621	
	Median		3.5000	
	Variance		.453	
	Std. Deviation		.67301	
	Minimum		1.00	
	Maximum		4.00	
	Range		3.00	
	Interquartile Range		.75	
	Skewness		-1.594	.285
	Kurtosis		2.851	.563
motivation scale	Mean		2.8756	.07680
	95% Confidence Interval for Mean	Lower Bound	2.7224	
		Upper Bound	3.0288	
	5% Trimmed Mean		2.8974	
	Median		3.0000	
	Variance		.419	
	Std. Deviation		.64717	
	Minimum		1.17	
	Maximum		4.00	
	Range		2.83	
	Interquartile Range		.83	
	Skewness		-.568	.285
	Kurtosis		-.112	.563

Note that skewness for *competence* is greater than |1.0|, which means skewness is higher than desirable

pleasure scale	Mean		3.1479	.07138
	95% Confidence Interval for Mean	Lower Bound	3.0055	
		Upper Bound	3.2902	
	5% Trimmed Mean		3.1800	
	Median		3.2500	
	Variance		.362	
	Std. Deviation		.60145	
	Minimum		1.50	
	Maximum		4.00	
	Range		2.50	
	Interquartile Range		.75	
	Skewness		-.737	.285
	Kurtosis		.127	.563

Minimum is 1.5 and maximum is 4.0, both of which are reasonable values for the pleasure variable

---

Interpretation of Output 2.5b

Note in your output that for *competence* and *motivation* two students each have missing data and the valid **N** is 71 (not shown here). Next, we check the normality assumption for variables that were labeled as scale to be sure that the variables truly are approximately normally distributed. Look at the skewness value in Output 2.2b to see if it is between −1 and 1. From the output we see that most of these variables have skewness values between −1 and 1, but competence at −1.59 is more skewed than is desirable. Thus, it may be helpful to change *competence* from scale to ordinal in the **Measure** column of the **Variable View**, and to use nonparametric statistics with it.

There are several ways to check this assumption in addition to checking the skewness value. If the mean, median, and mode, which can be obtained with the **Frequencies** command, are approximately equal, then you can assume that the distribution is approximately normally distributed. Although the mean and median for *competence* are not greatly different from each other (3.29 vs. 3.5), the median is outside of the confidence intervals for the mean (3.13 to 3.45), again suggesting that the distribution is slightly more skewed that is optimal, but not greatly skewed.

In addition to numerical methods for understanding your data, there are several graphical methods. SPSS can create histograms with or without the normal curve superimposed and also frequency polygons (line graphs) to roughly assess normality. You should use the plots to supplement, rather than to replace descriptive statistics because visual inspection can be deceiving because some approximately normal distributions may not look very much like a normal curve. Nevertheless, it can be useful to look at the plots to visualize the distribution. For this reason, we requested stem-and-leaf plots, histograms, and boxplots, which we will interpret here.

Output 2.5c: Stem and Leaf plot of Father's Education

father's education Stem-and-Leaf Plot

Frequency	Stem & Leaf
22.00	2 . 0000000000000000000000
16.00	3 . 0000000000000000
3.00	4 . 000
8.00	5 . 00000000
4.00	6 . 0000
1.00	7 . 0
7.00	8 . 0000000
6.00	9 . 000000
6.00	10 . 000000

Stem width: 1
Each leaf: 1 case(s)

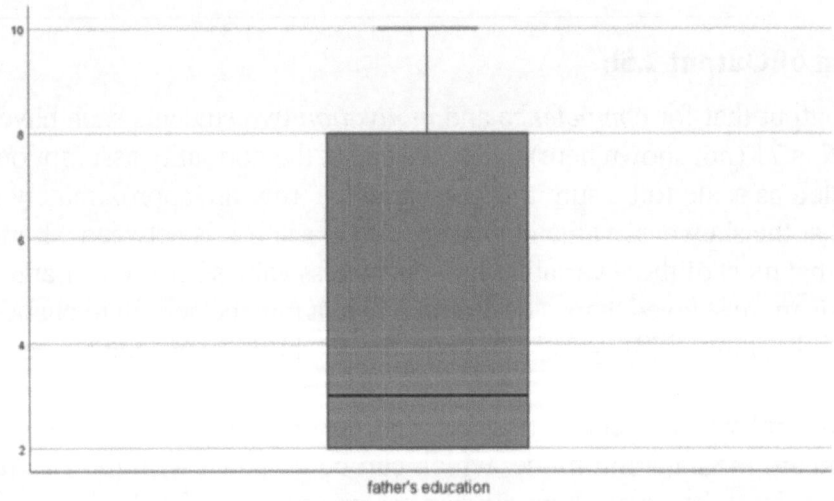

father's education

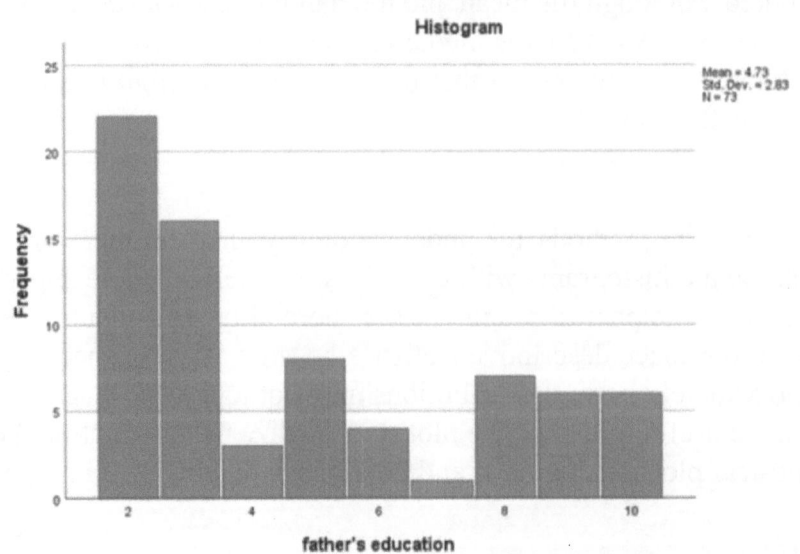

Histogram

Mean = 4.73
Std. Dev. = 2.83
N = 73

father's education

Interpretation of Plots in Output 2.5a, 2.5b, and 2.5c

Output 2.5a and 2.5b include stem-and-leaf plots, histograms, and boxplots, in addition to the statistics we already discussed. You may notice that the shape of the stem-and-leaf plot is the same as that of the histogram, but one is rotated 90 degrees compared to the other. Each of these plots gives you a visual of the distribution of the scores. The stem-and-leaf plot also includes numbers of cases with each score (Frequency) and the scores themselves (Stem & Leaf). Note that the legend at the bottom of the plot indicates the "stem width," which is 1 for *father's education* and 10 for *math achievement*. This means that each number in "Stem" column for *father's education* is a number between 2 and 10. The "Leaf" indicates the numbers in the next place (after the decimal point if the "Stem" is 1, or in the ones place if the "Stem" is 10). Thus, the *father's education* scores are all whole numbers, because the stem width is 1 and the "leaves" are all zeros. The Stem-and-Leaf plot thus gives you a lot of information in a single graph. However, you can see that although the plots may look somewhat skewed, the skewness statistic suggests that this amount of skewness is not enough to be considered a problem.

The second plot is the boxplot. The "box" represents the middle 50% of the cases, and the "whiskers" at the top and bottom of the box indicate the "expected" top and bottom 25%. The line running horizontally through the box represents the median. The fact that it is not located in the middle of the box is one indication that the variable is at least somewhat skewed. Another indicator is that the whisker on one end of the box is much longer than the whisker on the other end (which is not even visible). However, the fact that there are no **outliers** indicates that the skew is not extreme. This inference is supported by the fact that skewness was less than $|1.0|$. If there were **outliers**, there would be Os, and if there were really extreme scores, they would be shown with asterisks, above or below the end of the whiskers. Notice that there are not any Os or asterisks in the boxplot in Output 2.2a. So the boxplot, like the histogram and Stem-and-Leaf plot, indicate that there is some skewness to the data, but that it is not extreme. This shows why it is desirable to examine the skewness statistic as well. It is .684, which is less than 1.0, indicating that skewness is not sufficiently extreme that one needs to worry that results of most parametric tests would be inaccurate.

The third plot is a histogram. It provides a bigger picture of the frequency distribution, in the orientation with which most researchers are familiar.

Problem 2.6: Boxplots Split by a Dichotomous Variable

Many times, researchers are interested in reporting information on one specific variable for multiple subgroups of the participants (e.g., students who did and didn't take advanced math). Creating separate boxplots as well as separate statistics and stem-and-leaf plots can be useful if you want to see if the distributions of scores are very different for the two groups, which would suggest that the variances are unequal. It is also useful for checking to see if skew is in opposite directions for the groups, which can cause greater difficulties for parametric difference statistics if the variances are also unequal.

2.6. Is there heterogeneity of variances, and are the shapes of the distribution of *math achievement* approximately normal for both students who took algebra 2 and those who didn't take algebra 2?

To get Output 2.6:

- **Analyze → Descriptive Statistics → Explore**.
- The **Explore** window will appear.
- Click on *math achievement* and move it to the **Dependent List**.
- Next, click on *alg2* and move it to the **Factor** (or independent variable) **List**.
- Click on **Both** under **Display**. This will produce both a table of descriptive statistics and two kinds of plots: **Stem-and-Leaf** and **Box-and-Whiskers**.
- Click on **OK**.

Output 2.6: Boxplots Split by Taking/Not Taking Algebra 2 in With Statistics and Stem-and-Leaf Plots

```
EXAMINE VARIABLES=mathach BY alg2
  /PLOT BOXPLOT STEMLEAF
  /COMPARE GROUPS
  /STATISTICS DESCRIPTIVES
  /CINTERVAL 95
  /MISSING LISTWISE
  /NOTOTAL.
```

Case Processing Summary

| | algebra 2 in h.s. | Cases | | | | | |
| | | Valid | | Missing | | Total | |
		N	Percent	N	Percent	N	Percent
math achievement test	not taken	40	100.0%	0	0.0%	40	100.0%
	taken	35	100.0%	0	0.0%	35	100.0%

Descriptives

	algebra 2 in h.s.			Statistic	Std. Error
math achievement test	not taken	Mean		8.3917	.75190
		95% Confidence Interval for Mean	Lower Bound	6.8708	
			Upper Bound	9.9125	
		5% Trimmed Mean		8.3982	
		Median		7.8335	
		Variance		22.614	
		Std. Deviation		4.75546	
		Minimum		-1.67	

	Maximum		17.00	
	Range		18.67	
	Interquartile Range		7.75	
	Skewness		.102	.374
	Kurtosis		-.694	.733
taken	Mean		17.3333	.87834
	95% Confidence Interval for Mean	Lower Bound	15.5483	
		Upper Bound	19.1183	
	5% Trimmed Mean		17.6243	
	Median		18.3330	
	Variance		27.002	
	Std. Deviation		5.19631	
	Minimum		5.33	
	Maximum		23.67	
	Range		18.33	
	Interquartile Range		8.00	
	Skewness		-.543	.398
	Kurtosis		-.525	.778

Interpretation of Output 2.6

The first table under **Explore** provides descriptive statistics about the number of students who didn't take algebra 2 and the number of students who did take algebra 2 with **Valid** and **Missing** data. Note that we have 40 who didn't take algebra 2 and 35 who did take algebra 2 with valid *math achievement test* scores.

The **Descriptives** table contains many different statistics for students who didn't and did take algebra 2 separately. Note that the average *math achievement test* score is 8.39 for students who didn't take algebra 2 and 17.33 for those who did (over twice the score), and the maximum score for those who didn't take algebra 2 is lower than the mean score for those who did take algebra 2. Note that for both groups, the skewness values are less than (absolute value) 1, which indicates that *math achievement* **is approximately normal for both groups of students**. This is an assumption of the *t* test and ANOVA, and multivariate versions of this assumption are required for many of the statistics performed in this book.

The **Descriptives** table also provides the variances for each group. A key assumption of ANOVA and the *t* test is that the variances are approximately equal (i.e., the assumption of homogeneity of variances). Note that the variance is 22.61 for students who didn't take algebra 2 and 27.00 for those who did. These do not seem grossly different, and if we computed a Levene test on the differences in variances between students who took algebra 2 and students who did not take the course on this variable, we would find that the difference is not statistically significant. Thus, **the assumption of homogeneous variances is *not* violated**. The boxplots and stem-and-leaf plots will help you see this.

```
math achievement test Stem-and-Leaf Plot for
alg2= not taken

Frequency  Stem &    Leaf
     1.00    -0 .    1
     8.00     0 .    11233344
    17.00     0 .    55555667777899999
    11.00     1 .    00123334444
     3.00     1 .    777

Stem width: 10.0
Each leaf:      1 case(s)

math achievement test Stem-and-Leaf Plot for
alg2= taken

Frequency  Stem &    Leaf
     2.00     0 .    56
    11.00     1 .    00012444444
     9.00     1 .    557788999
    13.00     2 .    0111222333333

Stem width: 10.0
Each leaf:      1 case(s)
```

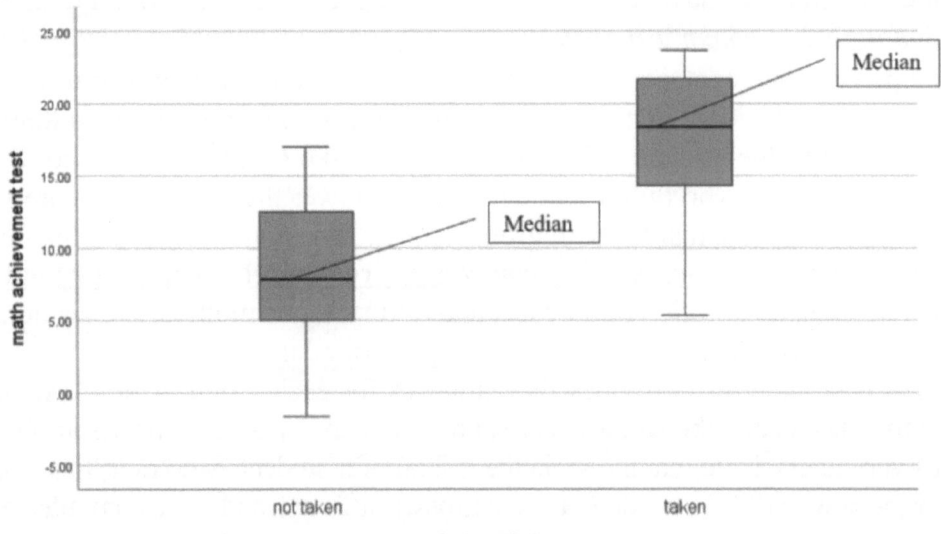

Interpretation of Output 2.6 continued

The **Stem-and-Leaf Plots**, for each group separately, are the next part of the output. As before, they give a visual impression of the distribution, and they show *each* person's score on the dependent variable (*math achievement*). As the legend indicates, **Stem width** equals 10. This means that entries that have 0 for the stem are less than 10, with the leaf indicating the actual number (1–9), those with 1 as the stem range from 10 to 19, etc. Note also that the legend indicates that each **leaf** equals one case. Each number in the **Leaf** column represents the last digit of one person's *math achievement* score. The numbers in the **Frequency** column indicate how many participants had scores in the range represented by that stem. Thus, in the "no algebra 2" plot, one student had a **stem** of –0 and a **leaf** of 1, that is, a score of –1. When you see this negative score, it alerts you to make sure this is a legitimate score. Look at the **variable view** for the SPSS spreadsheet to see the **Value Labels,** so you can see if scores can be negative. Can they? The plots also make it clear that the majority of students who took algebra 2 scored well on math achievement. About 2/3 of them made scores of 15 or higher; in contrast, only three students who didn't take algebra 2 made scores of 15 or higher.

Note also that the **Boxplots** are the last part of the output. There are two boxplots (one for each group). By inspecting the plots, we can see that the median score for those taking algebra 2 is quite a bit higher than that for those who didn't, with no overlap in the boxes. However, there is some overlap of the distributions. It is not clear whether the clearly better performance of students who took algebra 2 is because of algebra 2 content on the math achievement test, or is mainly because students who do not perform well in math are less likely to take more advanced math courses such as algebra 2. In later chapters, we will show you inferential statistics (analysis of covariance; mediation analysis) that could potentially tell us whether this difference is explained by differences in another variable (e.g., grades in math).

Using the output to check your data for errors. Checking the boxplots and stem-and-leaf plots can help identify outliers that might be data entry errors. In this case, there are none.

Using the output to check your data for assumptions. As noted in the interpretation of Outputs 2.2a and 2.2b, you can tell if a variable is grossly nonnormal by looking at the boxplots. The fact that the median is near the center of the box and the whiskers are reasonably symmetrical suggests an approximately normal distribution. The stem-and-leaf plots provide similar information. Importantly, skewness values also are well under $|1.0|$.

Problem 2.7: Using Tables for Data Description With Dichotomous Variables

Descriptives for Dichotomous Variables

We will now use the **Frequencies** command for each of the dichotomous variables. Once again, we could have done **Explore**, but we chose **Frequencies** because the standard deviation, variance, and skewness values are not very helpful with dichotomous variables, but frequencies are.

2.7.   What proportion of the sample has a value of "1," the higher value, for *algebra 1, algebra 2, geometry, trigonometry, calculus,* and *math grades*?

To produce Output 2.7:

- Select **Analyze → Descriptive Statistics → Frequencies**.

After selecting **Frequencies**, you will be ready to compute the *N*, minimum, maximum, and mean for all participants or cases on all selected variables in order to examine the data.

- Before starting this problem, press **Reset** to clear the **Variable** box if you have any variables in the box.
- While holding down the control key (i.e., "Ctrl") *highlight all* of the **dichotomous** variables in the left box. These variables have only two levels: *algebra 1, algebra 2, geometry, trigonometry, calculus, and math grades*.
- Click on the **arrow** button pointing right.
- Be sure that all of these variables have moved out of the left window and into the **Variable(s)** window.
- Click on **Statistics.**
- Select **Mean, Minimum**, and **Maximum**.
- Click on **Continue.**
- Click on **OK.**

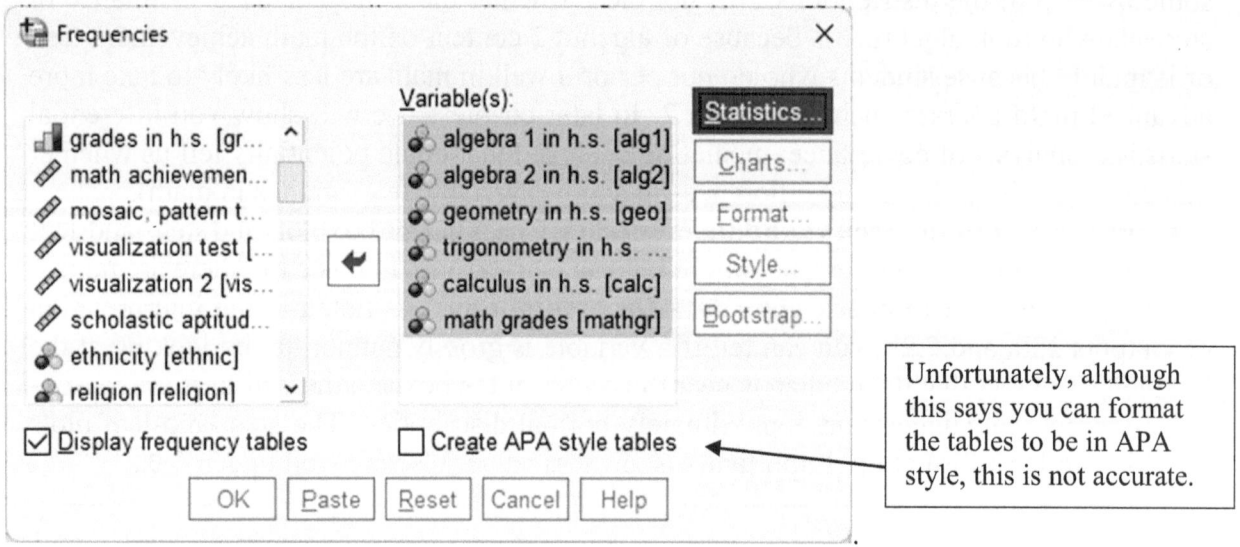

Figure 2.16 Frequencies.

Compare your output with Output 2.4

```
FREQUENCIES VARIABLES=alg1 alg2 geo trig calc mathgr
  /STATISTICS=MINIMUM MAXIMUM MEAN
  /ORDER=ANALYSIS.
```

This mean tells us that about 79% of students took Algebra 1 in high school (were coded as "1" rather than "0").

Statistics

		algebra 1 in h.s.	algebra 2 in h.s.	geometry in h.s.	trigonometry in h.s.	calculus in h.s.	math grades
N	Valid	75	75	75	75	75	75
	Missing	0	0	0	0	0	0
Mean		.79	.47	.48	.27	.11	.41
Minimum		0	0	0	0	0	0
Maximum		1	1	1	1	1	1

Frequency Table

More precisely, the frequency table tells us, 78.7% of students took Algebra 1 in high school. Compare the means and frequency table results for the other dichotomous variables as well.

algebra 1 in h.s.

		Frequency	Percent	Valid Percent	Cumulative Percent
Valid	not taken	16	21.3	21.3	21.3
	taken	59	78.7	78.7	100.0
	Total	75	100.0	100.0	

algebra 2 in h.s.

		Frequency	Percent	Valid Percent	Cumulative Percent
Valid	not taken	40	53.3	53.3	53.3
	taken	35	46.7	46.7	100.0
	Total	75	100.0	100.0	

geometry in h.s.

		Frequency	Percent	Valid Percent	Cumulative Percent
Valid	not taken	39	52.0	52.0	52.0
	taken	36	48.0	48.0	100.0
	Total	75	100.0	100.0	

trigonometry in h.s.

		Frequency	Percent	Valid Percent	Cumulative Percent
Valid	not taken	55	73.3	73.3	73.3
	taken	20	26.7	26.7	100.0
	Total	75	100.0	100.0	

calculus in h.s.

		Frequency	Percent	Valid Percent	Cumulative Percent
Valid	not taken	67	89.3	89.3	89.3
	taken	8	10.7	10.7	100.0
	Total	75	100.0	100.0	

math grades

		Frequency	Percent	Valid Percent	Cumulative Percent
Valid	less A-B	44	58.7	58.7	58.7
	most A-B	31	41.3	41.3	100.0
	Total	75	100.0	100.0	

Interpretation of Output 2.7

Output 2.7 includes only one table of **Descriptive Statistics**. Across the top row are the variable names, and the rows display the requested statistics of N, **Minimum**, **Maximum**, and **Mean**. We could have requested other statistics, but they would not be very meaningful for dichotomous variables. Down the left column are the variable labels. The N column indicates that all the variables have complete data. The **Valid N (listwise)** is 75, which also indicates that all the participants had data for each of our requested variables.

The most helpful column is the **Mean** column. Although the mean is not meaningful for nominal variables with more than two categories, you can use the mean of dichotomous variables to understand what percentage of participants fall into each of the two groups. For example, the mean of *academic track* is .55, which indicates that 55% of the participants were coded as 1 (regular track); thus, 45% were coded 0 (fast track). Because the mean is greater than .50, there are more regular track students than fast track students. If the mean is close to 1 or 0 (e.g., algebra 1 and calculus), then splitting the data on that dichotomous variable might not be useful because there will be many participants in one group and very few participants in the other.

> **Using your output to check your data for errors**. The **Minimum** column shows that all the dichotomous variables had "0" for a minimum, and the **Maximum** column indicates that all the variables have "1" for a maximum.

Problem 2.8: Using Frequency Tables for Multi-Category Variables

Displaying frequency tables for multi-category nominal or ordinal variables can help you understand how many participants are in each level of a variable and how much missing data of various types you have. For nominal variables, most descriptive statistics are meaningless. Thus, having a frequency table is usually the best way to understand your nominal variables, those with three or more unordered categories or levels. It can also be useful to understand ordinal variables with just a few levels. Unfortunately, **Frequencies** will not calculate interquartile range, which is the best measure of variability for ordinal variables. You can either calculate it by subtracting the 25th percentile from the 75th percentile or by running **Explore**.

2.8.   What is the frequency of each value for *ethnicity* (a nominal variable) and for *father's education* (an ordered variable)?

To produce Output 2.8:

- Select **Analyze → Descriptive Statistics → Frequencies**.
- Click on **Reset** if any variables are in the **Variable(s)** box.
- Now *highlight* the **nominal** variable, *ethnicity*, in the left box.
- Click on the **arrow** button pointing right.
- Highlight and move over the **ordinal** variable, *father's education*.
- Be sure the **Display frequency tables box** is checked.
- Do not click on **Statistics** because we do not want to select any this time.
- Click on **OK**.

Compare your output to Output 2.5.

Output 2.8 Frequency Tables for a Nominal Variable and an Ordinal Variable

```
FREQUENCIES VARIABLES=ethnic faed
  /ORDER=ANALYSIS.
```

Frequencies

Statistics

		ethnicity	father's education
N	Valid	73	73
	Missing	2	2

Frequency Table

As the prior table shows, two participants have missing data. The "Valid Percent" column tells you, for cases who are NOT missing data only, the percentage who are in each category. The percent column includes missing data in the denominator.

ethnicity

		Frequency	Percent	Valid Percent	Cumulative Percent
Valid	Euro-Amer	41	54.7	56.2	56.2
	African-Amer	15	20.0	20.5	76.7
	Latino-Amer	10	13.3	13.7	90.4
	Asian-Amer	7	9.3	9.6	100.0
	Total	73	97.3	100.0	
Missing	multi ethnic	1	1.3		
	blank	1	1.3		
	Total	2	2.7		
Total		75	100.0		

father's education

		Frequency	Percent	Valid Percent	Cumulative Percent
Valid	< h.s. grad	22	29.3	30.1	30.1
	h.s. grad	16	21.3	21.9	52.1
	< 2 yrs voc	3	4.0	4.1	56.2
	2 yrs voc	8	10.7	11.0	67.1
	< 2 yrs coll	4	5.3	5.5	72.6
	> 2 yrs coll	1	1.3	1.4	74.0
	coll grad	7	9.3	9.6	83.6
	master's	6	8.0	8.2	91.8
	MD/PhD	6	8.0	8.2	100.0
	Total	73	97.3	100.0	
Missing	System	2	2.7		
Total		75	100.0		

Cumulative Percent includes all prior categories, so it is really just appropriate for ordinal or interval variables. 74% of fathers have less than a bachelor's degree.

Interpretation of Output 2.8

There is one **Frequency** table for *ethnicity* and one for *father's education*. The left-hand column shows the **Valid** categories (or levels or values), **Missing** values, and **Total** number of participants. The **Frequency** column gives the number of participants who had each value. The **Percent** column is the percent who had each value, including missing values. For example, in the ethnicity table, 54.7% <u>**of all participants**</u> were *Euro-American*, 20.0% were *African-American*, 13.3% were *Latino-American*, and 9.3% were *Asian-American*. There also were a total of 2.7% missing: 1.3% were *multiethnic*, and 1.3% didn't answer so were left *blank*. The **valid percent** shows the percentage of those with *nonmissing* data at each value; for example, 56.2% of the 73 students <u>**with a single valid ethnic group**</u> were *Euro-Americans*. Finally, **Cumulative Percent** is the percentage of subjects in a category *plus* the categories listed above it. This last column is not very useful with nominal data such as ethnicity but can be quite informative for frequency distributions with several ordered categories. For example, in the distribution of father's education, 74% of the fathers had less than a bachelor's degree (i.e., they had not graduated from college).

    Using your output to check your data for errors. Errors can be found by checking to see if the number missing is the number you expected. Also, if you have more than one type of missing data and you assigned different numbers to these (e.g., 98 and 99), you will see the types listed in the first column.

    Using the output to check your data for assumptions. Frequency tables are helpful for checking the levels of the variable to see if you have subjects in each one. If one of the levels does not have many subjects in it, it can create problems with using that as an independent variable in difference statistics (see Chapters 3 and 7 for an explanation of types of difference statistics).

Describing the Sample Demographics and Key Variables

Using the information that you computed, we can now describe this sample of the High School and Beyond (hsb) data. In an article or research report such as a thesis, you will, at a minimum, state the number of participants and provide summary information about the age, gender, and ethnicity characteristics of the sample. We do not have age or gender as variables in this dataset; all the participants were high school seniors and are assumed to be about 18 years old, and we will write as though we had gender information. We do have *ethnicity*. For a more complete description of sample demographics, we will add mother's and father's education (revised). This information probably would be described in the *Participants* part of the Method section of an article or thesis and could include a table as shown in the box labeled **How to Write About Sample Demographics**. Note that there is always text preceding a table and that the text highlights and often simplifies the material in the table; <u>never repeat everything in the table or include a table without referring to it in the text</u>.

How to Write About Sample Demographics

Method

Participants

Data from 75 high school seniors (34 males and 41 females) were gathered. The majority of the group (54.7%) was of European-American ethnicity. Table 2.1 shows the frequencies and percentages of students by ethnicity and parent education. Note that data on education were missing for two fathers. Percentages for Father's Education refer to percentage of those for whom data are available. The majority of the mothers (64%) and fathers (52.1%) had a high school education or less. Approximately 26% of the fathers but less than 11% of the mothers had a bachelor's degree or more.

Table 2.1
Demographics of a Sample of 75 High School Seniors

Characteristic	*n*	%
Ethnicity		
European-American	41	54.7
African-American	15	20.0
Latino-American	10	13.3
Asian-American	7	9.3
Multi-ethnic	1	1.3
Missing	1	1.3
Mother's education		
H.S. grad or less	48	64.0
Some college	19	25.3
Bachelor's or more	8	10.7
Father's education		
H.S. grad or less	38	52.1
Some college	16	21.9
Bachelor's or more	19	26.0

The results section of a research report or thesis will often provide descriptive data and, perhaps, a table about the key variables in the study. In a research manuscript, you would actually place the tables at the end, after the reference list. This description of the variables may, but usually will not, directly address the research questions for the study. Several key variables, whether or not students had taken each of five types of mathematics courses and how they performed in all their math courses, were dichotomous. Other key variables, such as various test scores and attitudes about mathematics, were essentially continuous, and most were approximately normally distributed. In the box labeled **How to Write About Key Variables**, we show one way to describe the variables and summarize them in tables. The data for math courses taken and pleasure come from Output 2.6.

Note that means, standard deviations, and skewness are all rounded to *two* decimal places. Statistical symbols with English letters (e.g., *N*, *n*, *M*, *SD*) are italicized in the text and tables.

> Notice that the table is described in the text. In a journal manuscript, the tables are placed at the end, after the references, rather than immediately after the text about them. Otherwise, Table 2.2 is formatted in correct APA style.

How to Write About Key Variables

Results

Table 2.2 shows the means, standard deviations, and skewness for two key variables. The average student said that experiencing pleasure when solving math problems was typical of him/her (over 3.0 on a 4-point scale), with scores ranging from 1.5 (atypical/very atypical) to 4.0 (very typical). On average, students took 2.11 math courses, with the number taken ranging from none to five. Both variables were reasonably normal in distribution.

Table 2.2

> Notice that there are no vertical lines, the title is italicized, and there are horizontal lines above and below the column headers and after the final row of text and numbers but not in between rows.

Means, Standard Deviations, Skewness, Minimum, and Maximum for Key Variables

Variable	*M*	*SD*	*Skewness*	*Minimum*	*Maximum*
Pleasure scale	3.13	0.60	−0.68	1.5	4.0
Math courses taken	2.11	1.67	0.33	0.0	5.0

Problem 2.9: Bar Charts

With **nominal** data, you should not use a graphic that connects adjacent categories because with nominal data there is no necessary ordering of the categories or levels and the distribution is not continuous. Thus, it is better to make a bar graph or chart of the frequency distribution of variables like *religion, ethnic group,* or other nominal variables; the points that happen to be adjacent in your frequency distribution are not by necessity adjacent.

2.9. What are the relative numbers of high school students in this sample who report being Protestant, Catholic, and "no religion" (Almost no students identified with religions other than Christianity, and this included extremely small *N*s for any one religion, so they were treated as missing)?

To produce Output 2.9:

- **Select Analyze → Descriptive Statistics → Frequencies** . . . The **Frequencies** window should open.
- Click on *religion* and move it into the **Variable(s):** box.
- Click on **Charts** . . . This will open the **Frequencies: Charts** window.
- Select **Bar charts** and then **Continue**.
- Click off the check mark next to **Display frequency tables**. We do not need the frequency tables in this analysis.
- Click on **OK**. Compare your output with Output 2.9.

Output 2.9 Frequency Distribution Bar Chart for the Nominal Variable of Religion

```
FREQUENCIES VARIABLES=religion
  /FORMAT=NOTABLE
  /BARCHART FREQ
  /ORDER=ANALYSIS.
```

Frequencies

Statistics

religion

N	Valid	67
	Missing	8

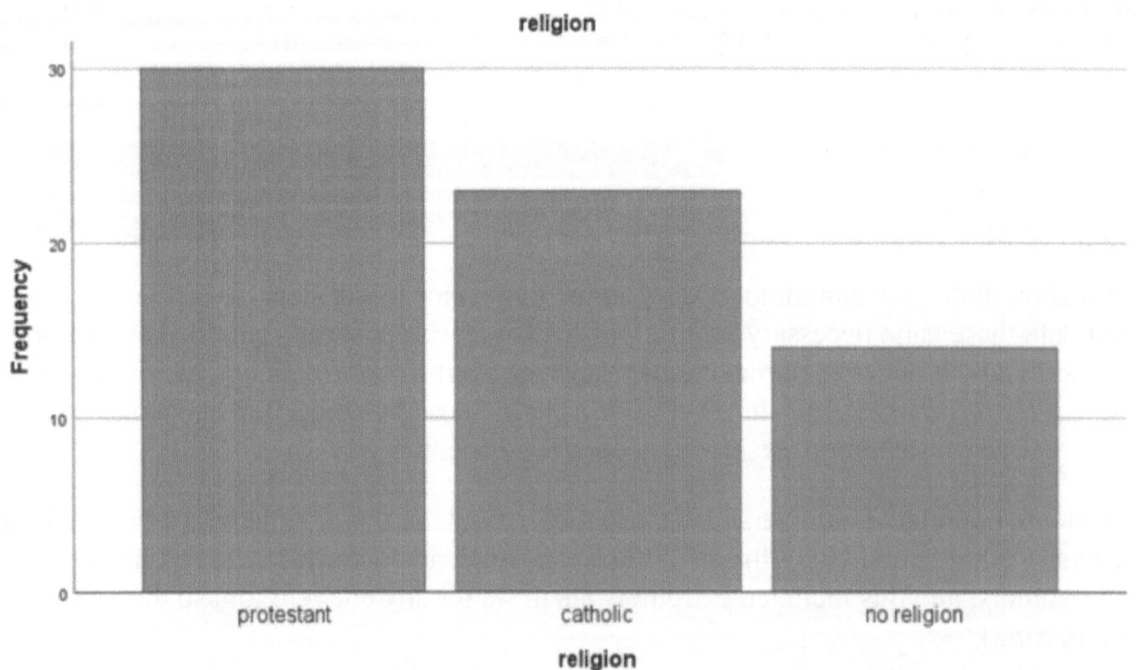

Interpretation of Output 2.9

There could be two parts to the output: a small statistics table and the bar chart. The full **Frequencies** table was not requested, since you saw this before. The **Statistics** table again indicates the valid *N* and the missing *N*. There are more missing data for this variable, probably because students with a religion other than Christianity were treated as missing, as were those who preferred not to answer. The **bar chart** presents a bar for each level of the nominal variable. Keep in mind that the order of the bars is arbitrary.

    Using your output to check your data for errors. Errors can be seen if the levels are not what you were expecting. For example, if you inadvertently entered the wrong number (a number that does not have a category label), you will see a bar for that number, with the number used as a label for the bar. Also, if there are no participants for a level, there will not be a bar for that level.

    Using the output to check your data for assumptions. Bar charts are helpful if you are unsure how many categories of a variable actually occur in your data; you can count the number of bars. You can also see the number of participants in each level. It is best if there are approximately the same number of subjects in each level if one wants to use the variable as an independent variable in procedures like ANOVA and MANOVA.

Table 2.3 summarizes which descriptive statistics and charts are most useful for each level of measurement.

Table 2.3

Descriptive Statistics and Charts to Use for Each Level of Measurement

	Nominal	*Dichotomous*	*Ordinal*	*Normal*
Frequency Distribution	Yes[a]	Yes	Yes	OK[b]
Bar Chart	Yes	Yes	Yes	OK
Histogram	No[c]	No	OK	Yes
Box and Whiskers Plot	No	No	Yes	Yes
Central Tendency				
Mean	No	OK	Of ranks, OK	Yes
Median	No	OK = Mode	Yes	OK
Mode	Yes	Yes	OK	OK
Variability				
Range	No	Always = 1	OK	Yes
Standard Deviation	No	No	Of ranks, OK	Yes
Interquartile Range	No	No	Yes	OK
How Many Categories	Yes	Always = 2	OK	Not if truly continuous
Shape				
Skewness	No	No	Yes	Yes

[a]Yes = This is recommended [b]OK = Although this is not best practice, it is not truly wrong [c]No = Should not use

Interpretation Questions

2.1. Using Outputs 2.5a and 2.5b: (a) What is the mean *visualization test* score? (b) What is the range for *grades in h.s.?* (c) What is the minimum score for *mosaic pattern test?* How does this compare with the values for that variable as indicated in Chapter 1? Why could the minimum be a negative number?

2.2. Using Outputs 2.5b: (a) For which of the variables that we called scale is the skewness statistic more than +/−1.00? (b) Why is the answer important? (c) How many participants have some missing data? (Hint: Check Chapter 1 if you don't remember the sample size.) (d) What percent of students have a valid (nonmissing) *motivation* or *competence* score? (e) Can you tell from Output 2.1b how many are missing both *motivation* and *competence* scores?

2.3 Using Output 2.7: (a) Can you interpret the means? Explain. (b) How many participants are there altogether? (c) How many have complete data (nothing missing)? (d) What percent took *algebra 1*?

2.4. Using Output 2.8: (a) 9.6% of what set of participants are *Asian-American*? (b) What percent of students have fathers who had a high school education or less? (c) What percent of fathers (with a known education) have a master's degree or higher?

2.5. (a) Compare and contrast nominal, dichotomous, ordinal, and normal variables. (b) In social science research, why isn't it important to distinguish between interval and ratio variables?

2.6. What percent of the area under the standard normal curve is within one standard deviation of (above or below) the mean? What does this tell you about scores that are more than one standard deviation away from the mean?

2.7. (a) How do z scores relate to the normal curve? (b) How would you interpret a z score of −3.0? (c) What percentage of scores is between a z of −2 and a z of +2? Why is this important?

Extra SPSS Problems

Using the *college student data.sav* file, do the following problems. Print your outputs and circle the key parts of the output that you discuss.

2.1. For the variables with five or more ordered levels, compute the skewness. Describe the results. Which variables in the dataset are approximately normally distributed/scale? Which ones are ordered but not normal, and can the normality of these variables be improved by transforming them?

2.2. Do a stem-and-leaf plot for *same-sex parent's height* split by *academic track*? Discuss the plots.

2.3. Which variables are nominal? Run frequencies for the nominal variables and other variables with fewer than five levels. Comment on the results.

2.4. Do boxplots for *student height* and for *hours of study*. Compare the two plots.

2.5. Create frequency polygons of your choice. Describe the meaning of the frequency polygon.

2.6. You ran descriptive statistics on two continuous variables. One variable (V1) had a skewness of 1.34 and the second variable (V2) had a skewness of .45. Explain why this difference is meaningful.

2.7. Using data of interest to you or your instructor, run descriptive analysis of that dataset and write up a two-page analysis of those data.

3 Selecting and Interpreting Inferential Statistics

To understand the information in this chapter, it is important to remember or to review the sections in Chapter 1 about **variables** and **levels of measurement** (nominal, dichotomous, ordinal, and normal/scale). It is also necessary to remember the distinction we made between **difference** and **associational** research questions and between **descriptive** and **inferential statistics**. This chapter focuses on inferential statistics, which, as the name implies, refers to statistics that make inferences about population values based on the sample data that you have collected and analyzed. **Difference inferential statistics** lead to inferences about the differences (usually mean differences) between groups in the populations from which the samples were drawn. **Associational inferential statistics** lead to inferences about the association or relationship between variables in the population. Thus, the purpose of inferential statistics is to enable the researcher to make generalizations beyond the specific sample data. Before we describe how to select statistics, we introduce design classifications.

General Design Classifications for Difference Questions

Many research questions focus on whether there is a significant difference between two or more groups or conditions, with those groups/conditions defined by a categorical independent variable (also known as a grouping variable or factor). When a group comparison or difference question is asked, the independent variable and design can be classified either as between groups or within subjects. Understanding this distinction is one essential aspect of determining the proper statistical analysis for this type of question.

Labeling difference question designs. It is helpful to have a brief descriptive label that identifies the design for other researchers and also guides us toward the proper statistics to use. There are no comparable design classifications for the descriptive or associational research questions, so <u>this section only applies to difference questions</u> (questions involving comparisons of two or more groups or conditions in their relative amount of a dependent variable). Designs are usually labeled in terms of (a) whether the members of each of the groups being compared are independent of those in the other groups (between groups) versus the people in each group/condition are not independent of one another or are the same people (dependent samples or within subjects), (b) the number of independent variables, and (c) the number of levels within each independent variable. The dependent variable is not part of the label of the design.

DOI: 10.4324/9781003355908-3

Between-groups designs. These are designs where each participant in the study is in one and only one condition or group and members of one group are independent of those in all other groups. For example, in a study investigating the "effects" of fathers' education on *math achievement*, there may be three groups (or levels or values) of the independent variable, *father's education*. These levels are: (a) *high school or less*, (b) *some college*, and (c) *bachelor's degree or more*. In a between-groups design, each participant is in only one of the three conditions or levels and the level of *math achievement* of students in these three groups is compared. If the investigator wished to have 20 participants in each group, then 60 participants would be needed to carry out the research.

Within-subjects designs. These designs are conceptually the opposite of between-groups designs. In within-subjects designs (also called **repeated measures**, or **related samples** designs), the participants in each of the conditions or levels of the independent variable are somehow connected to those in the other conditions or levels of the independent variable. Often, this is because the same participants receive or experience all of the conditions or are assessed on the dependent variable at each of the times at which these assessments occur; however, these designs also include examples where the participants are matched by the experimenter or are related in some other way (e.g., twins, husband and wife, or mother and child). In that case, each type of person (e.g., husband vs. wife or child with developmental disability vs. mental-age matched comparison child) is one level of the independent variable in a **related samples** design. When each participant is assessed on the same measure more than once, within-subjects designs are also referred to as **repeated measures** designs. Repeated measures designs are common in longitudinal research and intervention research. Comparing performance on the same dependent variable assessed before and after intervention (pretest and posttest) is a common example of a repeated measures design. We might call the independent variable in such a study "time of measurement" or "change over time." In the HSB dataset, one of the variables is repeated (*visualization score* with two levels, *visualization* and *visualization 2*), and one is related samples (*education*, each student has both a *mother's education* and *father's education*). We will use a paired or matched statistic to see if *mother's education* is on the average higher or lower than *father's education*.

Single-factor designs. If the design has only one independent variable (either a between-groups design or a within-subjects design), then it should be described as a single-factor or one-way design. **Factor** and **way** are other names for group difference independent variables. Note that the number of factors or "ways" refers to the number of *independent variables*, not the number of *levels/groups* of an independent variable. For example, a between-groups design with one independent variable that has four levels is a single-factor or "one-way" between-groups design with four levels. If the design was a within-subjects design with four levels, then it would be described as a single-factor repeated measures design with four levels (e.g., the same test being given four times).

Between-groups factorial designs. When there is more than one group difference independent variable, and each level of each independent variable is possible in combination with each level of the other independent variables, the design is called **factorial**. For example, a factorial design could have two independent variables (i.e., factors), *academic track* and *ethnicity*, allowing for fast and regular track students of each ethnic group. In these cases, the number of levels of *each* factor (independent variable) becomes important in the description of the design. Each independent variable (IV) is described using one number, which refers to the number of levels for that variable. For example, if *academic track* has two levels (fast and regular) and *ethnicity* has three levels (Euro-American, African-American, and Latinx-American), then this design is a 2 × 3 (read "two by three") between-groups factorial design. In this way of describing the design, the *number* of numbers (in this case two, the 2 and the 3) is the number of factors or ways, and the *numbers themselves* refer to the number of levels of each of those factors. This design could also be called a two-way or two-factor design because there are two independent variables.

Mixed factorial designs. If the design has at least one between-groups variable and at least one within-subjects independent variable, it is called a **mixed design**. For example, let's say that the two independent variables are academic track (a between-groups variable with 2 levels—fast and regular) and time of measurement (with pretest and posttest as the two within-subjects levels); this is a 2 × 2 mixed factorial design with repeated measures on the second factor. The mixed design is common in experimental studies with a pretest and post-test, but the analysis can be complex so it is discussed in our Intermediate SPSS book (Barrett et al., 2025).

So to summarize, when describing a design, each independent variable (IV) is described using one number, which is the number of levels for that variable. Thus, a design description with two numbers (e.g., 3 × 4) has two independent variables or factors, which have three and four levels. The dependent variable (DV) is not part of the design description, so it was not addressed in these design descriptions.

Selection of Inferential Statistics

How do you decide which of the many possible inferential statistics to use? Although this section may seem overwhelming at first because many statistical tests are introduced, don't be concerned if you don't now know much about the tests mentioned. You should come back to this chapter later, when you have to make a decision about which statistic to use, and by then, the tests will be more familiar. We present eight steps, shown in Figure 3.1, to help guide you in the selection of an appropriate statistical test. The steps and tables are our recommendations; you will see there are often other appropriate choices.

Remember that **difference questions** compare groups and utilize the statistics that we call difference inferential statistics. These statistics (e.g., *t* test and ANOVA) are shown in Tables 3.1 and 3.3 and are discussed in the chapters noted in the tables.

Associational questions utilize what we call associational inferential statistics. The statistics in this group examine the association or relationship between two or more variables and are shown in Tables 3.2 and 3.4.

Using Tables 3.1 to 3.4 to Select Inferential Statistics

As with research questions and hypotheses discussed in Chapter 1, we divide inferential statistics into basic and complex. For **basic (or bivariate) statistics**, there is *one* independent and *one* dependent variable (Table 3.1) or two variables that may or may not be viewed as independent and dependent variables (Table 3.2). For **complex statistics**, there are three or more variables. We decided to call them **complex** rather than **multivariate**, which is more common in the literature,

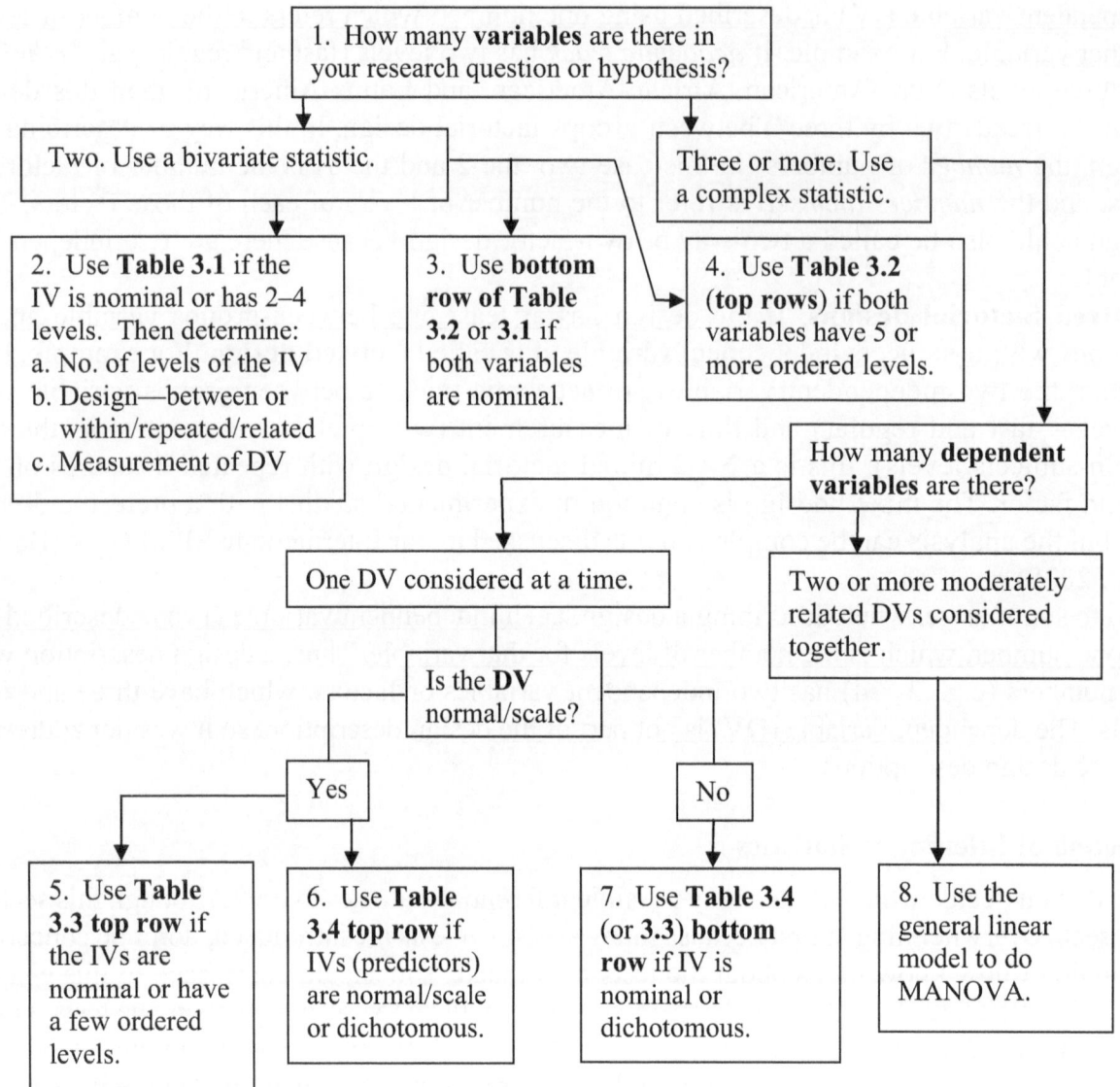

Figure 3.1 A decision tree to help you select the appropriate inferential statistic from Tables 3.1 to 3.4 (IV = independent variable; DV = dependent variable).

because there is not unanimity about the definition of multivariate, and some statistics with more than two variables (e.g., factorial ANOVA) are not usually classified as multivariate. For complex statistics, you will use Table 3.3 or 3.4. Most of the statistics shown in Tables 3.1 and 3.2 are discussed in the remaining chapters in this book, and text is provided demonstrating how to compute and interpret them.

Two of the complex statistics in Tables 3.3 and 3.4 (**factorial ANOVA** and **multiple regression**) are introduced in this book, but they and other such complex statistics are discussed in more detail in our *IBM SPSS for Intermediate Statistics* book (Barrett et al., 2025). These statistics are identified with I.B. (i.e., intermediate book) in the tables. These four tables include most of the inferential statistics that you will encounter in reading research articles. Note the boxes in the decision tree are numbered to correspond to the numbers in the text here, which expands on the decision tree.

1. Decide <u>how many variables</u> there are in your research question or hypothesis. If there are only two variables, use Table 3.1 or 3.2. If there is *more* than one independent and/or dependent variable (i.e., three or more variables) to be used in this analysis, use Table 3.3 or 3.4.

Basic (Two Variable) Statistics

2. If the <u>independent variable is nominal</u> (i.e., has unordered levels) **or** <u>has a few (2–4) ordered levels</u>, use Table 3.1. Then your question is a **basic** two variable **difference question** to compare groups. You must then determine: (a) whether there are <u>two *or* more than two levels</u> (also called categories or groups or samples) of your independent variable, (b) whether the design is <u>between groups or within subjects</u>, and (c) whether the <u>measurement level</u> of the dependent variable is (i) <u>normal/scale</u> and parametric assumptions are not markedly violated *or* (ii) <u>ordinal</u> *or* (iii) <u>nominal</u> *or* dichotomous (see Chapter 2 if you need help). The answers to these questions lead to a specific box and statistic in Table 3.1.

For example, if there are (a) three levels of the independent variable, if (b) the design is between groups (i.e., these are different unrelated participants in the three groups), and if (c) the dependent variable is normally distributed, the appropriate statistic is **one-way ANOVA** (analysis of variance), which is shown in the top row, second box from the right, and discussed in Chapter 7. Another example could be that (a) the independent variable has two levels, (b) the design is within-subjects (repeated measures), and (c) the dependent variable data are ordinal or markedly violate the parent's assumptions. In this case, the appropriate statistic would be the **Wilcoxon** test shown in Table 3.1. An example of the selection of an **independent samples *t* test** is presented at the end of this chapter. Examples of selection, computation, and interpretation of all the statistics noted with "Ch. x" in Tables 3.1–3.4 are discussed in the referenced chapter.

3. If <u>both variables are nominal or dichotomous</u>, you could ask <u>either a difference question</u> (use the bottom row of Table 3.1; e.g., **chi-square**) <u>or an associational question</u> and use the bottom row of Table 3.2 to select **phi** or **Cramer's *V***. Note, in the second to bottom row of Table 3.2, we have included **eta**, an associational statistic used with one nominal and one normal or scale variable. Later we will see it used as an effect size measure with ANOVAs. There are many other nonparametric associational measures, some of which we will see in Chapters 7 and 8, as indicated in these two Tables 3.1 and 3.2.

4. If <u>both variables have many (we suggest five or more) ordered levels</u>, use Table 3.2 (top two rows). Your research question would be a **basic** two variable (bivariate) **associational question**. Which row you use depends on *both* variables. If both are normal/scale, then you would probably select the **Pearson product moment correlation** or **bivariate regression** (top row). Note that **bivariate** simply means there are only two variables involved (one IV and one DV). This is to distinguish bivariate from multiple regression, which you will more often read about in the scientific literature, but all of the analyses in Tables 3.1 and 3.2 are bivariate; the type of analysis is **regression.** Regression should be used if one has a clearly directional hypothesis, with an independent and dependent variable. Correlation is chosen if one is simply

Table 3.1

Selection of an Appropriate Inferential Statistic for Bivariate (Two Variable), Difference Questions or Hypotheses[a]

Scale of Measurement of **Dependent Variable**	Appropriate **Descriptive** Statistics to report	One Factor or Independent Variable With **Two Levels** or **Categories**/Groups/Samples		One Independent Variable **Three or More Levels** or Groups	
		Independent Samples or Groups **(Between)**	Repeated Measures or Related Samples **(Within)**	Independent Samples or Groups **(Between)**	Repeated Measures or Related Samples **(Within)**
Dependent Variable Approximates **Normal/Scale** Data and Assumptions Not Markedly Violated	Mean, Standard Deviation	INDEPENDENT SAMPLES *t* TEST Ch. 7	PAIRED SAMPLES *t* TEST Ch. 7	ONE-WAY ANOVA Ch. 7	GLM REPEATED MEASURES ANOVA I.B.[b]
Dependent Variables **Clearly Ordinal** or Parametric Assumptions Markedly Violated	Median (or Mean Rank), Interquartile Range	MANN-WHITNEY Ch. 7	WILCOXON Ch. 7	KRUSKAL-WALLIS Ch. 7	FRIEDMAN I.B.[b]
Dependent Variable **Nominal** or **Dichotomous**	Mode, Minimum & Maximum	CHI-SQUARE Ch. 5	MCNEMAR	CHI-SQUARE Ch. 5	COCHRAN Q TEST

[a] It is acceptable to use statistics that are in the box(es) below the appropriate statistic, but there is usually some loss of power. It is not acceptable to use statistics in boxes above the appropriate statistic or ones in another column.

[b] I.B. = Barrett et al. (2025) *IBM SPSS for Intermediate Statistics: Use and Interpretation* (6th ed.). The other chapters mentioned in the table are the chapters where those statistics are discussed in this *IBM SPSS for Introductory Statistics* book.

Table 3.2

Selection of an Appropriate Inferential Statistic for Bivariate (Two Variable), Associational Questions or Hypotheses

Measurement of **Both Variables**	Relate	Two Variables or Scores for the Same or Related Subjects
Variables Are Both **Normal/Scale** and Assumptions Not Markedly Violated	SCORES	PEARSON (r) or BIVARIATE REGRESSION Ch. 6
Both Variables at Least **Ordinal** Data or Assumptions Markedly Violated	RANKS	KENDALL'S TAU-B or SPEARMAN (Rho) Ch. 6
One Variable Is **Normal/Scale** and One Is **Nominal**		ETA Ch. 5
Both Variables Are **Nominal** or **Dichotomous**	COUNTS OF PEOPLE	PHI or CRAMER'S *V* Ch. 5

interested in how the two variables are related. If one or both variables are ordinal (ranks or grossly skewed) or other assumptions are markedly violated, the second row (**Kendall's tau** or **Spearman rho**) is a better choice.

Complex (Three or More Variable) Questions and Statistics

It is possible to break down a complex research problem or question into a series of basic (bivariate) questions and analyses. However, there are advantages to combining them into one complex analysis; additional information is provided and a more accurate overall picture of the relationships is obtained.

5. If you have one normally distributed (scale) dependent variable (DV) and two (or perhaps three or four) independent variables (IVs), each of which is nominal or has a few (2–4) ordered levels, you will use the top row of Table 3.3 and one of three types of **factorial ANOVA**. These analysis of variance (ANOVA) statistics answer **complex difference questions**.

Many of these tests are beyond the scope of this book, but are discussed in our companion, intermediate book. **Log-linear** analysis is a nonparametric statistic somewhat similar to the between-groups factorial ANOVA for the case where all the variables are nominal or dichotomous (see Table 3.3). The programs Mixed Models and Nonparametric Statistics provide some useful tests if your dependent variable is ordinal.

6. The statistics in Table 3.4 are used to answer **complex associational questions**. If you have two or more independent or predictor variables and one normal (scale) dependent variable, the top row of Table 3.4 and **multiple regression** are appropriate.
7. If the dependent variable is dichotomous or nominal, consult the bottom row of Table 3.4, and use **discriminant analysis** or **logistic regression**, both discussed in Barrett et al. (2025).
8. Multivariate analysis of variance, **MANOVA**, is used if you have two or more normal (scale) dependent variables treated simultaneously. This complex statistic is discussed in our intermediate SPSS book (Barrett et al., 2025).

Table 3.3

Selection of the Appropriate Complex (Two or More Independent Variables) Statistic to Answer Difference Questions or Hypotheses

Dependent Variable(s)	Two or More Independent Variables		
	All Between Groups	All Within Subjects	Mixed (Between & Within)
One **Normal/Scale** Dependent Variable	GLM, Factorial ANOVA or ANCOVA Ch. 7 and I.B.[a]	GLM (ANOVA) Repeated Measures on All Factors I.B.	GLM (ANOVA) Repeated Measures on Some Factors I.B.
Ordinal Dependent Variable	Generalized Linear Models	Generalized Estimating Equations	Generalized Estimating Equations
Dichotomous Dependent Variable	Log Linear; Generalized Linear Models	Generalized Estimating Equations	Generalized Estimating Equations

[a] I.B. = Barrett et al. (2025) *IBM SPSS for Intermediate Statistics: Use and Interpretation* (6th ed.).

Table 3.4

Selection of the Appropriate Complex Associational Statistic for Predicting a Single Dependent/Outcome Variable From Several Independent Variables

One Dependent or Outcome Variable	Several Independent or Predictor Variables		
	Normal or Scale	Some Normal Some Dichotomous (Two category)	All Dichotomous
Normal/Scale (Continuous)	MULTIPLE REGRESSION Ch. 6 and I.B.[a]	MULTIPLE REGRESSION Ch. 6 and I.B.	MULTIPLE REGRESSION Ch. 6 and I.B.
Dichotomous	DISCRIMINANT ANALYSIS I.B.	LOGISTIC REGRESSION I.B.	LOGISTIC REGRESSION I.B.

[a] I.B. = Barrett et al. (2025) *IBM SPSS for Intermediate Statistics: Use and Interpretation* (6th ed.).

Exceptions

Occasionally you will see a research article in which a dichotomous *dependent variable* was used with a *t* test or ANOVA, or as either variable in a Pearson correlation. Because of the special nature of dichotomous variables, this is not necessarily wrong, as would be the use of a nominal (three or more unordered levels) dependent variable with these parametric statistics. However, we think that it is usually a better practice to use the same statistics with dichotomous variables that you would use with nominal variables. The exception is that it is appropriate to use dichotomous (dummy) independent variables in multiple and logistic regression (see Table 3.4 again).

The General Linear Model

Whether or not there is a relationship between variables can be answered in two ways. For example, if each of two variables provides approximately normally distributed data with five or more levels, based on Figure 3.1 and Table 3.2, the statistic to use is either the Pearson correlation or bivariate (two variable) regression, and that would be our recommendation. Instead, some researchers choose to divide the independent variable into two or more levels or groups, such as low, medium, and high, and then do a one-way ANOVA.

Conversely, in a second example, others who start with an independent variable that has only a few (say two through four *ordered* categories) may choose to do a correlation instead of a one-way ANOVA. Although these choices are not necessarily wrong, we do not think they are usually the best practice. In the first example, information is lost by dividing a continuous independent variable into a few categories. In the second example, there would be a restricted range, which tends to decrease the size of the correlation coefficient.

In the previous examples, we recommended one of the choices, but the fact that there *are* two choices raises a bigger and more complex issue. Statisticians point out, and can prove mathematically, that the distinction between difference and associational statistics is an artificial one. Figure 3.2 (and Chapter 1, Figure 1.1) shows that, although we make a distinction between difference and associational inferential statistics, they both serve the purpose of examining (top box) relationships and both are subsumed by the general linear model (middle box).

Statisticians state that all common parametric statistics are relational. Thus, the full range of methods used to analyze the linear relationship between one continuous dependent variable and one or more independent variables, either continuous or categorical, can be computed using the same mathematical equation. The model on which this is based is called the **general linear model**. The relationship between the independent and dependent variables can be expressed by an equation with weights for each of the independent/predictor variables plus an error term. Difference questions are computed using one or more dichotomous predictors indicating group membership.

The bottom part of Figure 3.2 indicates that a *t* test or one-way ANOVA with one dichotomous independent variable is analogous to bivariate or simple regression. Finally, as shown in the

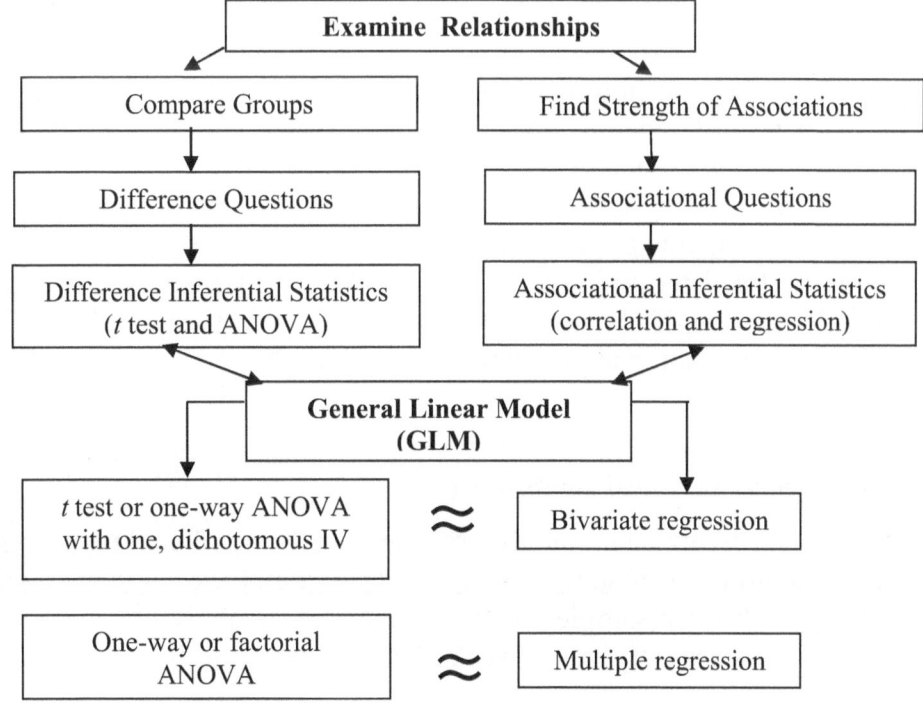

Figure 3.2 A general linear model diagram of the selection of inferential statistics.

lowest row of boxes in Figure 3.2, a one-way or factorial ANOVA can be computed mathematically as a multiple regression with multiple dichotomous predictors (dummy variables). Note in Figure 3.1 and Tables 3.1 and 3.3 that SPSS uses the GLM program to perform a variety of statistics, including factorial ANOVA and MANOVA.

Although we recognize that our distinction between difference and associational parametric statistics is a simplification, we think it is useful because it corresponds to a distinction made by most researchers when they are conceptualizing their studies. We hope that this glimpse of an advanced topic is clear and helpful.

Interpreting the Results of a Statistical Test

In Chapters 5, 6, and 7, we present information about how to check assumptions, do the commands, interpret the previous statistics, and write about them. For each statistic, the program produces a number or **calculated value** based on the specific data in your study. They are labeled t, F, and so on, or sometimes just value by SPSS.

Statistical Significance

The calculated value is compared to a **critical value** that takes into account the degrees of freedom, which are usually based on the number of participants. Using this comparison, the program provides a probability value (usually called p value in articles, but usually called *sig.* for significance in SPSS). It is the probability that results of that magnitude, with a sample of that size, is due to chance. Researchers want to guard against Type I error, which is the probability of rejecting the null hypothesis when it is actually true, so they decide how much Type I error is permissible, by setting the p value that is allowable at alpha (usually .05). Figure 3.3 shows how to interpret any inferential test once you know the probability level (p or *sig.*) from the computer or statistics table. In general, if the calculated value of the statistic (t, F, etc.) is relatively large and the sample size is also large, the p value will be small (e.g., .05, .01, .001). If the probability is *less than* the preset alpha level, we can say that the results are **statistically significant**, or that they are significant at the .05 level if alpha = .05. We can also say that the null hypothesis of no difference or no relationship can be rejected. Note that, using this program's computer printouts, it is quite easy to determine statistical significance because the actual significance or probability level (p) is printed so you do not have to look up a critical value in a table. SPSS labels this p value as **Sig.**, and all of the common inferential statistics have a common metric for determining significance, this p value or **Sig**. Thus, regardless of what specific statistic you use, if the sig. or p is small (e.g., less than an alpha set at .05), the finding is statistically significant, and you can reject the null hypothesis of no difference or no relationship.

Practical Significance versus Statistical Significance

Students, and sometimes researchers, misinterpret statistically significant results as being practically or clinically important. But statistical significance is not the same as practical significance or importance. With large samples, you can find statistical significance even when the differences or associations are very small/weak. Thus, in addition to statistical significance,

SPSS Value[1]	Meaning	Null Hypothesis	Interpretation
1.00	$p = 1.00$	Fail to Reject	Not Statistically Significant (could be due to chance)
.50	$p = .50$		
.06	$p = .06$		
.049	$p < .05$ or $p = .049$	Reject[2]	Statistically Significant[3] (not likely due to chance but not necessarily important or practical)
.01	$p = .01$		
.000	$p < .001$		

1. SPSS uses **Sig.** to indicate the significance or probability level (p) of all inferential statistics. This figure provides just a sample of Sig. values, which could be any value from 0 to 1.
2. .05 is the typical alpha level that researchers use to assess whether the null hypothesis should be rejected or not. However, occasionally, researchers use more liberal levels (e.g., .10) in exploratory studies or, more commonly, more conservative levels (e.g., .01) if more than one statistical test is being run.
3. Statistically significant does *not* mean that the results have practical significance or importance. See next section.

Figure 3.3 Interpreting inferential statistics using Sig.

you should consider confidence intervals and examine effect size. We will see that it is quite possible, with a large sample, to have a statistically significant result that is weak (i.e., has a small **effect size**). Remember that the null hypothesis states that there is *no* difference or *no* association. A statistically significant result with a small effect size means that we can be very confident that there is at least a little difference or association, but that relationship may not be of any practical importance.

Confidence Intervals

One alternative or addition to null hypothesis significance testing (NHST) is creating confidence intervals. These intervals provide more information than NHST and *may* provide more practical information. If we conducted the same study an infinite number of times, then in 95% of those studies, the true population value would fall within the 95% confidence interval. This can give us more specific information about practical significance. For example, suppose one knew that an increase in reading scores of five points, obtained on a particular instrument, would lead to a functional increase in reading performance. Then, reading scores for students experiencing two different methods of instruction were compared. The result showed that students who used the new method scored significantly higher statistically than those who used the other method. According to NHST, we would reject the null hypothesis of no difference between methods and conclude that the new method is better. If we apply **confidence intervals** to this same study, we can be confident that the *population mean difference* lies between the upper and lower bounds 95% of the times the study is repeated and confidence intervals calculated. If the lower bound of that interval is greater than five points, we can conclude that it is likely that using this method of instruction would lead to a practical or functional increase in reading levels. If, however, the confidence interval ranged from, say, 1 to 11, the result would be statistically significant, but the mean difference in the population could be as little as 1 point, or as big as 11 points. Given these results, we could not be confident that there would be a *practical* increase in reading using the new method. If the confidence interval includes 0 (zero), we fail to reject the null hypothesis.

Effect Size

A statistically significant outcome does not give information about the strength or size of the outcome. Therefore, it is important to know, in addition to information on statistical significance, the size of the effect. **Effect size** is defined as the strength of the relationship between the independent variable and the dependent variable, and/or the magnitude of the difference between levels of the independent variable with respect to the dependent variable. Statisticians have proposed many effect size measures that fall mainly into two types or families, the *r* family and the *d* family.

   The *r* family of effect size measures. One method of expressing effect sizes is in terms of strength of association or *relationship*. The most well-known variant of this approach is the **Pearson correlation coefficient, *r***. Using Pearson *r*, effect sizes always have an absolute value less than 1.0, varying between −1.0 and +1.0 with 0 representing no effect and +1 or −1 the maximum effect. This *family* of effect sizes includes many other associational statistics such as Spearman's rho (r_s), phi (Φ), eta (η), and the multiple correlation (R).

   The *d* family of effect size measures. The *d* family focuses on magnitude of *difference* rather than on strength of association. If one compares two groups using a *t* test, SPSS will calculate the effect size (*d*) by subtracting the mean of the second group (B) from the mean of the first group (A) and dividing by the pooled standard deviation of the two groups. So *d* tells you the extent to which the difference between the two means in level of the outcome variable is greater than standard/regular variability in the variable. The general formula is on the left. If the two groups have equal *n*s, the pooled *SD* is the average of the *SD*s for the two groups. When *n*s are unequal, the formula on the right is the appropriate one. If you need to calculate *d* to compare two means as a follow-up to ANOVA, be sure to follow the rules of algebra, doing multiplications before addition. The answer to the denominator is the SD weighted more for the larger sample; it should be a number between the SDs of the two groups.

$$d = \frac{MA - MB}{SD_{pooled}} \qquad\qquad d = \frac{M_A - M_B}{\sqrt{\dfrac{(N_A - 1)SD_A^2 + (N_B - 1)SD_B^2}{N_A + N_B - 2}}}$$

There are many other formulas for *d* family effect sizes, but they all express effect size in standard deviation units. Thus, a *d* of .5 means that the groups differ by one half of a standard deviation.

Using d, effect sizes usually vary from 0 to + or −1 but d can be more than 1 if the effect is very large.

Issues about effect size measures. Unfortunately, as just indicated, there are many different effect size measures and little agreement about which to use. Although d is the most commonly discussed effect size measure for differences between two groups, it is not available on many SPSS outputs. Only the **Compare Means and Proportions** program's **Independent Samples t test** and **Paired t test** provide d as the measure of effect size. If you use the **General Linear Model** programs, or even the **One-Way ANOVA** option in the **Compare Means and Proportions** program, SPSS does not calculate d as a measure of effect size. However, d can be calculated by hand from information in the printout, using the previous appropriate formula. The correlation coefficient, r, and other measures of the strength of association, such as phi (Φ), eta^2 (η^2), and R^2, are available in the other outputs if requested.

There is disagreement among researchers about whether it is best to express effect size as the unsquared or squared r family statistic (e.g., r or r^2). The squared versions have been used because they indicate the percentage of variance in the dependent variable that can be predicted from the independent variable (i.e., the explained variance). However, some statisticians argue that these usually small percentages give you an underestimated impression of the strength or importance of the effect. Thus, we (like Cohen, 1988) decided to use the unsquared statistics (r, Φ, η, and R) as our r family indexes.

There is also disagreement about whether or not to report effect size when results are not significant. The usual reason you would not have significant results even though you have a medium sized or large effect size is that your sample size is small. Therefore, you should be careful in interpreting the effect size for nonsignificant results. It is important not to assume that an effect size based on a small sample represents the population effect size. However, we believe it is still useful to report the effect size when results are not significant. If the effect size is extremely small, it is unlikely that it would be worthwhile to replicate the study with a larger sample, but it does seem worthwhile to do such a replication if the effect size is medium or large in size. This is particularly true when the observed p value (sig.) is close to .05.

Relatively few researchers reported effect sizes before 1999 when the APA Task Force on Statistical Inference stated that effect sizes should *always* be reported for your primary results (Wilkinson & APA Task Force, 1999). The 5th edition of APA adopted this recommendation of the Task Force so, now, most journal articles discuss the size of the effect as well as whether or not the result was statistically significant.

Interpreting Effect Sizes

Assuming that you have computed an effect size measure, how should it be interpreted? Based on Cohen (1988) and Vaske et al. (2002), Table 3.5 provides guidelines for interpreting the size of the "effect" for five common effect size measures: d, r, Φ, R and η.

Note that these guidelines are based on the effect sizes usually found in studies in the behavioral sciences and education. Thus, they do not have absolute meaning; large, medium, and small are only relative to typical findings in these areas. For that reason, we suggest using "larger than typical" instead of large, "typical" instead of medium, and "smaller than typical" instead of small. The guidelines do not apply to all subfields in the behavioral sciences, and they definitely do not apply to fields, designs, or contexts where the usually expected effects are either larger or smaller. It is advisable to examine the research literature to see if there is information about typical effect sizes on the topic and adjust the values that are considered typical accordingly.

Cohen (1988) provided research examples of what he labeled small, medium, and large effects to support the suggested *d* and *r* family values. Most researchers would not consider a correlation (*r*) of .5 to be very strong because only 25% of the variance in the dependent variable is predicted. However, Cohen argued that a *d* of .8 (and an *r* of .5, which he showed are mathematically similar) are "grossly perceptible and therefore large differences, as (for example, is) the mean difference in height between 13- and 18-year-old girls" (p. 7). Cohen stated that a small effect may be difficult to detect, perhaps because it is in a less well-controlled area of research. Cohen's medium size effect is "visible to the naked eye. That is, in the course of normal experiences, one would become aware of an average difference in IQ between clerical and semi-skilled workers" (p. 6).

Effect size and practical significance. The effect size indicates the strength of the relationship or magnitude of the difference and thus is relevant to the issue of practical significance. Although some researchers consider effect size measures to be an index of practical significance, we think that even <u>effect size measures are not direct indexes of the importance of a finding</u>. As implied earlier, what constitutes a large or important effect depends on the specific area studied, the context, and the methods used. Furthermore, practical significance always involves a judgment by the researcher and/or the consumers (e.g., clinicians, clients, teachers, school boards) of research that takes into account such factors as cost and political considerations. A common example is that the effect size of taking daily aspirin on heart attacks is quite small, but the practical importance is high because preventing heart attacks is a life or death matter, the cost of aspirin is low, and side effects are relatively uncommon (although recent research suggests that benefits exceed risk only for people who already had cardiovascular difficulties). On the other hand, a

Table 3.5

Interpretation of the Strength of a Relationship (Effect Sizes)

General Interpretation of the Strength of a Relationship	The d Family[a]	The r Family[b]					
	d	r^2	*r, rho, Φ*	R^2	*R*	η^2	*η (eta)*[d]
Much larger than typical	≥ |1.00|[c, e]	.49	≥ |.70|	.49+	|.70|+	.21	|.45|+
Large or larger than typical	|.80|	.25	|.50|	.26	|.51|	.14	|.37|
Medium or typical	|.50|	.09	|.30|	.13	|.36|	.06	|.24|
Small or smaller than typical	|.20|	.01	|.10|	.02	|.14|	.01	|.10|

[a] *d* values can vary from 0.0 to + or − infinity, but *d* greater than 1.0 is relatively uncommon.

[b] *r* family values can vary from 0.0 to + or −1.0, but except for reliability (i.e., same concept measured twice), *r* is rarely above .70. In fact, some of these statistics (e.g., phi) have a restricted range in certain cases; that is, the maximum phi (Φ) is less than 1.0.

[c] We interpret the numbers in this table as a range of values. For example, a *d* greater than .90 (or less than −.90) would be described as "much larger than typical," a *d* between, say, .70 and .90 would be called "larger than typical," and *d* between, say, .60 and .70 would be "typical to larger than typical." We interpret the other three columns similarly.

[d] Partial etas are multivariate tests equivalent to *R*. Use *R* column.

[e] Note. \| \| indicates absolute value of the coefficient. The absolute magnitude of the coefficient, rather than its sign, is the information that is relevant to effect size. *R* and η usually are calculated by taking the square root of a squared value, so that the sign usually is positive.

curriculum change could have a large effect size but be judged impractical because of high costs and/or extensive opposition to its implementation.

Confidence intervals of the effect size. Knowing the confidence interval around an effect size can provide information useful to making a decision about practical significance or importance. If the confidence interval is narrow, one would know that the effect size in the population is close to the computed effect size. On the other hand, if the confidence interval is large (as is usually the case with small samples) the population effect size could fall within a wide range, making it difficult to interpret the computed effect size for purposes of estimating practical significance. Similarly to the example described earlier, if the lower bound of the confidence interval was more than a minimum effect size agreed to indicate a practically significant effect, one could then be quite confident that the effect was important or practical. Unfortunately, SPSS does not provide confidence intervals for effect size measures for some tests (such as **General Linear Model)** and it is not easy to compute them by hand. It does provide confidence intervals for the **Compare Means and Proportions** program.

Steps in Interpreting Inferential Statistics

When you interpret inferential statistics, we recommend:

1. <u>Decide whether to reject the null hypothesis</u>. However, that is not enough for a full interpretation. If you find that the outcome is statistically significant, you need to answer at least two *more* questions. Figure 3.4 summarizes the steps described later about how to more fully interpret the results of an inferential statistic.

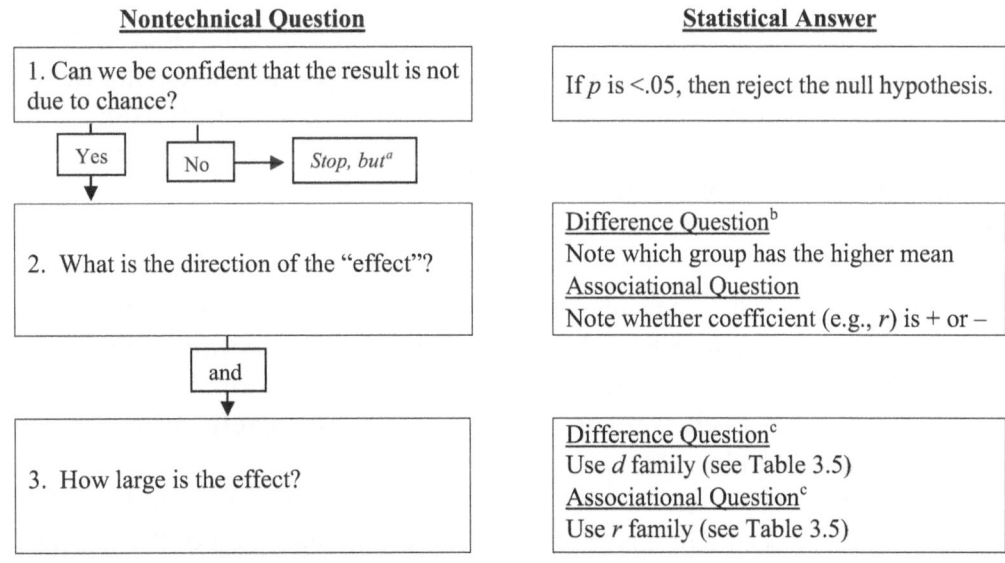

[a] If you have a small sample *(N)*, it is possible to have a nonsignificant result (it may be due to chance) and yet a large effect size. If so, an attempt to replicate the study with a larger sample may be justified.

[b] If there are three or more means or a significant interaction, a post hoc test (e.g., Tukey) will be necessary for complete interpretation.

[c] Interpretation of effect size is based on Cohen (1988) and Table 3.5. A "large" effect is one that Cohen stated is "grossly perceptible." It is larger than typically found but does not necessarily explain a large amount of variance. You might use confidence intervals in addition to or instead of effect sizes.

Figure 3.4 Steps in the interpretation of an inferential statistic.

2. <u>What is the direction of the effect?</u> Difference inferential statistics compare groups so it is necessary to state which group performed better. We discuss how to do this in Chapter 7. For associational inferential statistics (e.g., correlation), the sign is very important, so you must indicate whether the association or relationship is positive or negative. We discuss how to interpret associational statistics in Chapter 6.

3. <u>What is the size of the effect?</u> It is important to include the effect size and desirable to include the confidence interval in the description of all your results. Unfortunately, SPSS does not always provide the effect size measures you most want to report and/or confidence intervals for the tests and effect sizes, so for some statistics we may wish to compute or estimate the effect size by hand, or use an effect size calculator, several of which are available online.

4. **Although not shown in** Figure 3.4<u>, the researcher or the consumer of the research should also make a judgment about whether the result has practical or clinical significance or importance</u>. To do so, they need to take into account the effect size, the costs of implementing change, and the probability and severity of any side effects or unintended consequences.

An Example of How to Select and Interpret Inferential Statistics

As a review of what you read in Chapter 1 and this chapter, we now provide an extended example based on the HSB data. We will walk you through the process of identifying the variables, research questions, and approach, and then show how we selected appropriate statistics and interpreted the results.

Research problem. <u>Suppose your research problem was to investigate the relation of *academic track* and *math courses taken* to *math achievement*</u>.

Identification of the variables and their measurement. The research problem specifies three variables: *acadtrac*, *math courses taken*, and *math achievement*. The latter appears to be the outcome or dependent variable, and academic track and math courses taken are the independent or predictor variables because they occurred before the math achievement exam, which we said was taken near the end of the students' senior year. As such, *acadtrac* and *math courses taken* are *presumed* to have an effect on math achievement scores.

What is the level of measurement for these three variables? *Acadtrac* is clearly dichotomous (*fast* or *regular*). *Math courses taken* has six ordered values, from zero to five courses. These are scale (approximately normal) data because we found in Chapter 2 that the scores were not highly skewed. Likewise, the *math achievement* test has many levels, with more scores somewhere in the middle than at the high and low ends so was approximately normally distributed. We confirmed in Chapter 2 that math achievement was at least approximately normally distributed by determining the skewness was less than 1.

Research questions. There are a number of possible research questions that could be asked and statistics that could be used with these three variables, including all of the types of questions in Appendix B, the descriptive statistics discussed in Chapter 2, and several of the inferential statistics presented in this chapter. However, we focus on three research questions and three inferential statistics because they answer this research problem and fit our earlier recommendations for good choices. First, we discuss two basic research questions, given the previous specification of the variables and their measurement. Then, we discuss a complex research question that could be asked instead of research questions 1 and 2.

1. Is there a difference between fast track and regular track students (the two levels of the variable, *acadtrac*) on their average *math achievement scores*?

Type of research question. Using the text, Figure 3.1, and Table 3.1, you should see that the first question is phrased as a **basic difference question** because there are only two variables and the focus is a group difference (the difference between the fast track group and the regular track group).

Selection of an appropriate statistic. If you examine Table 3.1, you will see that the first question should be answered with an **independent samples *t* test** because (a) the independent variable has only two values (fast and regular), (b) the design is between groups (fast track and regular track students form two independent groups), and (c) the dependent variable (*math achievement*) is normal/scale data. We would also check other assumptions of the *t* test, especially homogeneity of variances, to be sure that they are not markedly violated.

Interpretation of the results for question 1. Let's assume that about 50 students participated in the study and that $t = 2.05$. The output will give you the exact **Sig.** In this case, <u>$p < .05$ and thus *t* is statistically significant. However, if you had 25 participants, this *t* would not have been significant</u> (because the *t* value necessary for statistical significance is influenced strongly by sample size. Small samples require a larger *t* to be significant.).

Deciding whether the statistic is significant only means the result is unlikely to be due to chance. You still have to state the direction of the result and the effect size and/or the confidence interval (see Figure 3.4). <u>To determine the direction, we need to know the mean (average) *math achievement* scores for fast track and regular track students</u>. In the HSB data, fast track students have the higher mean, as you will see in Chapter 7. Given that the difference is significant, you can be quite confident that fast track students in the population are at least a little better at math achievement, on average, than regular track students. So you should state that fast track students scored higher than regular track students. If the difference was not statistically significant, it is best *not* to make any comment about which mean was higher because the difference could be due to chance. Likewise, if the difference was not significant, we recommend that you do not interpret the effect size.[1] You should also provide the means and standard deviations so that the effect size can be better understood.

Because the *t* was statistically significant, we would <u>calculate *d* and discuss the **effect size**</u>, as shown earlier. In this situation, SPSS would compute the pooled (weighted average) standard deviation for fast track and regular track students' *math achievement* scores. Let's say that the difference between the means was 2.0 and the pooled standard deviation was 6.0; then *d* would be .33, a small to medium size effect. This means that the difference is less than typical of the statistically significant findings in the behavioral sciences. A *d* of .33 may or may not be a large enough difference to use for recommending programmatic changes (i.e., may or may not be practically significant).

Confidence intervals might help you decide if the difference in *math achievement* scores was large enough to have practical significance. For example, say you found (from the lower bound of the confidence interval) that you could only be confident that there was a ½-point difference between fast track and regular track students. Then you could decide whether that was a big enough difference to justify, for example, a programmatic change.

2. Is there an association between *math courses taken* and *math achievement*?

Type of research question. This second question is phrased as a **basic associational question** because there are only two variables, and both have many ordered levels. Thus, use Table 3.2 for the second question.

Selection of an appropriate statistic. As you can see from Table 3.2, the second research question should be answered with a **Pearson correlation** because both *math courses taken* and *math achievement* are normally distributed variables.

Interpretation of the results for research question 2. The interpretation of r is based on decisions similar to those made above for t. If $r = .30$ (with 50 subjects), it would be statistically significant at the $p < .05$ level. If the r is statistically significant, you still need to discuss the direction of the correlation and effect size. Because the correlation is positive, we would say that the *more* math courses students took, the *higher* they performed on the math achievement test and, conversely, the *fewer* math courses taken, the more poorly students performed on the math achievement test. The **effect size** of $r = .30$ is medium or typical. (In contrast, if the correlation were negative, this would mean that the more math courses a student took, the more poorly they performed on the math achievement course. A negative correlation does NOT mean that the variables are not related to one another).

Note that if N were 25, the r of .30 would not be significant, but the effect size (.30) would still be typical. On the other hand, if N were 500 and $r = .30$, p would be $<.0001$. With $N = 500$, even $r = .10$ would be statistically significant, indicating that you could be quite sure the association was not zero, but the effect size would be small, or less than typical.

Complex research question and statistics. There are advantages to considering the two independent variables (*academic track* and *math courses taken*) together rather than separately as in questions 1 and 2. There are several statistics that you will compute that could be used to consider *academic track* and *math courses taken* together, but many of these are beyond the scope of this book. A research question which subsumes both questions 1 and 2 could be:

3. Is there a combination of *academic track* and *math courses* that predicts *math achievement*?

Selection of an appropriate statistic. Multiple regression could be used to answer this question. As you can see in Table 3.4, multiple regression is appropriate because we are trying to predict a normally distributed/scale variable (*math achievement*) from two independent variables, which are *math courses taken* (normal or scale) and *academic track* (a dichotomous variable).

Based on our discussion of the general linear model (GLM), a **two-way factorial ANOVA** would be another statistic that could be used to consider both *academic track* and *math courses taken* simultaneously. However, to use ANOVA, the many levels of math courses taken would have to be recoded into a few categories or levels (perhaps high, medium, and low). Because information is lost when you do such a recode, we would not recommend factorial ANOVA for this example. Another possible statistic to use for this example is **analysis of covariance** (ANCOVA) using academic track as the independent variable and *math courses taken* as the covariate; ANCOVA is discussed in Barrett et al. (2025).

Interpretation of the results for research question 3. We provide an introduction to multiple regression in Chapter 6 and to factorial ANOVA in Chapter 7, but extended treatment is beyond the scope of this book (see Barrett et al., 2025). For now, let's just say that we would obtain important new information about the relationships among these three variables by doing these complex statistics rather than by doing only the *t* test and correlation described earlier and in the next section. In the HSB data, we would find that the fact that fast track students take more math courses than do regular track students seems to statistically account for the difference in math achievement of fast versus regular track students. Findings like this, which clarify potentially misleading results, are one of the reasons that complex statistics can be useful.

Writing About Your Outputs

One of the goals of this book is to help you write a research report or thesis using the outputs. Thus, we have provided an example later that could be two paragraphs from a research paper based on the expanded HSB data used in the assignments in this book.

Before demonstrating how you might write about the results of research questions 1 and 2, we would like to make several important points. There are several books listed in the "For Further Reading" that will help you write a research paper and make appropriate tables. Note especially the APA manual (2020) and Nicol and Pexman (2010a, 2010b). The example that follows and the samples provided in each output interpretation section give only one way to write about outputs. There are other good ways.

Based on your outputs, you should first write about descriptive statistics about the demographics (e.g., academic track, age, ethnicity) of the participants in your sample in your **Method** section. You should also include basic literature-based evidence about the reliability and validity of your measures or instruments and/or any such evidence based on your sample in the **Method** section. You also should include in your report whether statistical assumptions of the inferential statistics were met or how adjustments were made in an "Analytic Approach" section of the **Method** section, or if your analyses are straightforward and that section is not needed, then you should describe how you met the assumptions at the beginning of the results section in which you describe that analysis.

The **Results** section includes a description (but not a discussion) of the findings in words, tables, and maybe figures. Your Results section should include the following numbers about each statistically significant finding (in a table or the text):

1. The value of the statistic (e.g., $t = 2.05$ or $r = .30$) to two decimals.
2. The degrees of freedom (often in parentheses) and for chi-square the N (e.g., $\chi^2 = 5.26$, $df = 2$, $N = 49$).
3. The p or Sig. value (e.g., $p = .048$), preferably the exact p value to 2 or 3 decimal places.

4. If the statistic is significant, the direction of the finding (e.g., by showing which mean is larger or the sign of the correlation).
5. An index of effect size from either the *d* family or the *r* family, if the statistic is significant, and a statement about the relative size of the "effect." (See Table 3.5 and accompanying text.)

When not shown in a table, the prior information should be provided in the text, as shown later. *In addition* to the numerical information, you must <u>describe your significant results in words, including the variables related, the direction of the finding, and an interpretive statement about the size/strength of the effect</u> based on Table 3.5 or, better still, based on the effect sizes found in the literature on your topic. Realize that our effect size terms are only rough estimates of the magnitude of the "effect" based on what is typical in the behavioral sciences; they are not necessarily applicable to your topic.

If your paper includes a table, it is usually not necessary or advisable to include all the details about the value of the statistic, degrees of freedom, and *p* in the text because they are in the table. <u>If you have a table, you must refer to it by number (e.g., Table 1) in the text</u> and you must <u>describe the main points, but don't repeat all of it</u> or the table is not necessary. You can mention relationships that are not significant (acknowledging their statistical nonsignificance), but do not discuss the direction of the finding or interpret the meaning of nonsignificant findings because the results could well be due to chance. Do provide effect sizes or at least the information (e.g., *n*s, means, and standard deviations) necessary for other researchers to compute the effect size, in case your study is included in a future meta-analysis.

The **Discussion** section puts the findings in context in regard to the research literature, theory, and the purposes of the study. You should also discuss limitations of the study and attempt to explain why the results turned out the way they did.

How to Write About Results

Based on what we reported earlier about the results of research questions 1 and 2, we might make the following statements in our Results section:

Results

For research question 1, there was a statistically significant difference between fast track and regular track students on math achievement, $t(48) = 2.05$, $p = .04$, $d = .33$. Fast track students ($M = 14.70$) scored higher than regular track students ($M = 12.70$), and the effect size was small to medium according to Cohen's (1988) guidelines. The confidence interval for the difference between the means was .50 to 6.50, indicating that the difference could be as little as half a point, which is probably not a practically important difference, but also could be as large as six and one half points.

For research question 2, there was a statistically significant positive correlation between math courses taken and math achievement, $r(48) = .30$, $p = .03$. The positive correlation means that, in general, students who took more math courses tended to score higher on the math achievement test and students who took fewer math courses scored lower on math achievement. The effect size of $r = .30$ is considered medium or typical.

We present examples of how to write about the results of each statistic that you compute in the appropriate chapter.

Conclusion

Now you should be ready to study each of the statistics in Tables 3.1 to 3.4 and learn more about their computation and interpretation. It may be tough going at times, but hopefully this overview has given you a good foundation. It would be wise for you to review this chapter, especially the tables and figures, from time to time as you learn about the various statistical methods. If you do, you will have a good grasp of how the various statistics fit together, when to use them, and how to interpret the results. You will need this information to understand the chapters that follow.

Interpretation Questions

3.1. Compare and contrast a between-groups design and a within-subjects design.

3.2. What information about variables, levels, and design should you keep in mind in order to choose an appropriate statistic?

3.3. Provide an example of a study, including the variables, level of measurement, and hypotheses, for which a researcher could appropriately choose two different statistics to examine the relations between the same variables. Explain your answer.

3.4. When $p < .05$, what does this signify?

3.5. Interpret the following related to effect size:

(a) $d = .25$ (c) $R = .53$ (e) $d = 1.15$
(b) $r = .35$ (d) $r = .13$ (f) $\eta = .38$

3.6. What statistic would you use if you wanted to see if there was a difference between three ethnic groups on math achievement? Why?

3.7. What statistic would you use if you had two independent variables, income group ($<\$10,000$, $\$10,000–\$30,000$, $> \$30,000$) and ethnic group (Hispanic, Caucasian, African-American), and one normally distributed dependent variable (self-efficacy at work)?

3.8. What statistic would you use if you had one independent variable, geographic location (North, South, East, West), and one dependent variable (satisfaction with living environment, Yes or No)?

3.9. What statistic would you use if you had three normally distributed (scale) independent variables (weight of participants, age of participants, and height of participants), plus one dichotomous independent variable (academic track)?

3.10. A teacher *ranked* the students in her Algebra I class from 1 = highest to 25 = lowest in terms of their grades on several tests. After the next semester, she checked the school records to see what the students received from their Algebra II teacher. The research question is "Is there a relationship between rank in Algebra I and grades in Algebra II?" What statistic should she use?

3.11. Results of a t-test reported a difference between the stress levels of university students who had children compared to university students who did not have children. Using the Perceived Stress Score (PSS), university students with children averaged 7.2 and the stress level for university students without children was 5.6. The results of the t-test were t = 3.52, p = .032, and d = .84. Using the steps in Figure 3.4 write in interpretation of these results.

Note

1 At times when one has a small sample, one would present the effect size and talk about the need to replicate the study with a larger sample.

4 Methods to Provide Evidence for Reliability and Validity

In this chapter we start with a brief overview of measurement reliability and validity. Then we demonstrate four statistics that provide evidence for some of the several aspects of the reliability and validity of your data. These statistics are often described in the **Method** chapter of a research report because they precede the testing of your research hypothesis. Including information about the reliability and validity of your data will help support the answers to your research questions.

Problem 4.1 demonstrates the use of **Cohen's kappa**, which is a good method to assess evidence for interrater *reliability* of a nominal variable. Problem 4.2 uses **correlation** and the **paired samples *t* test** to illustrate one method to assess interrater *reliability* for normally distributed (scale) variables. Problem 4.3 uses **factor analysis** to show how one could reduce several variables to a smaller number of composite variables that represent several theoretically related aspects of a complex concept and provide support for later analyses of the research questions. In Problem 4.4, **Cronbach's alpha** is used to provide evidence for the internal consistency reliability of composite, multi-item subscales, or of sets of variables resulting from a factor analysis or from a theoretical combination of variables (e.g., several items on a questionnaire designed to measure the same concept). This chapter illustrates how to compute, interpret, and write about these statistics in terms of the evidence they provide to support reliability and validity.

Problems 4.1 and/or 4.2 will be helpful when you are using observations to "code" specific categories of behavior or whenever there is at least some subjectivity to a classification system. In this case the consistency or reliability of the data needs to be checked by having two observers or raters independently record their scores and then check how well they agree with one another (their interrater or interobserver reliability).

It is common for a researcher to develop a smaller number of new summated variables from an initially larger number of items such as the 14 Likert-type ratings we designed to measure attitudes about mathematics motivation, math competence, and pleasure doing math. The statistics described in Problems 4.3 and 4.4 are relatively complex but useful for assessing both the reliability and validity of established instruments such as published questionnaires. Even with an established questionnaire, one should check one's own data, at least for the internal consistency reliability (alpha) of the subconcepts such as the math attitudes of competence, motivation, and pleasure described in Chapter 1.

If one were developing a new or modified instrument, one would probably first try to find support for the validity of the theoretical groupings of the items to see how well the factor analysis (Problem 4.3) fits the proposed structure or organization of the subconstructs based on the literature or theory. Then, one would check the reliability of the factors using Cronbach's alpha, as we have in Problem 4.4.

DOI: 10.4324/9781003355908-4

It is important to realize that the statistics performed here are based on, and in some cases, the same as, those in Chapters 5–7. The Pearson correlation which was introduced briefly in Chapter 3 (especially in Table 3.2 and research question 2) is used here in Problems 4.2–4.4 so it is important to understand a little more about correlation before we introduce the problems in this chapter. (Correlation is discussed more fully in Chapter 6.)

Correlation coefficients. These statistics can vary from –1.00 (a perfect negative correlation or association) through .00 (no correlation) to +1.00 (a perfect positive correlation). Note that +1 and –1 are equally high or strong, but they lead to different interpretations. A high *positive correlation* between two items on a questionnaire would mean that persons with high ratings on the first item tended to have high ratings on the second item, those with lower item ratings on item one had lower item two ratings, and those in between had ratings that were neither especially high nor especially low. On the other hand, a high *negative correlation* would mean that students rated high on one item tended to rate low on the other item. With a *zero correlation* there is no consistent association. A person rated high on one item might be rated low, medium, or high on the other item.

Measurement Reliability

It is important to assess the reliability of your data prior to conducting the inferential statistics that help answer your research questions. If reliability tests indicate that your data have low reliability, the results from inferential testing would be suspect. To increase the chances that your new data will be reliable, it is best, whenever possible, to use existing measures that have already been tested. Regardless, it is always important to assess the degree of measurement reliability for your dataset. We use the term measurement reliability to point out that the data only provide evidence for reliability; it is inappropriate to say that a test or measure is "reliable" or "unreliable." Research methods books, such as Gliner et al. (2017), provide extended discussions of reliability and validity and the several methods used to provide evidence to support them.

Interrater Reliability

Reliability for one nominal variable assessed twice. There are several methods of computing interrater or interobserver reliability. In Problem 4.1, **Cohen's kappa** is used to assess interobserver agreement when the data are nominal or unordered categories. In this chapter we check the reliability of the *ethnicity* measure by examining the agreement between codes in the school records and ethnicity self-reported by the students.

Reliability for one ordered variable measured twice. In Problem 4.2, you will compute a **correlation coefficient** (as one part of a paired sample *t*) to check the interrater reliability of *visualization and mosaic scores*.[1] As mentioned earlier, ***interrater reliability*** is needed because behaviors or classification systems involve some degree of subjective judgment (e.g., when there are open-ended questions that must be classified, or ratings or categories observers must determine based on observations). If there were more than two raters the appropriate statistic would be the ICC (intraclass correlation coefficient), which could also be used with two raters. ICC is not demonstrated in this chapter; however, it is in our intermediate book, Barrett et al. (2025).

Internal Consistency Reliability

In Problem 4.4, we will compute the most commonly used type of reliability, **Cronbach's alpha**. This measure indicates the internal consistency of a multiple-item scale. Alpha is typically used when you have several Likert-type items that are summed to make a composite score or **summated scale**. Alpha is based on the mean or average correlation of each item in the scale with every other item. In the social science literature, alpha is widely used, in part because it provides a measure of reliability that can be obtained <u>during the study</u> from just one testing session or one administration of a questionnaire. In Problem 4.4, you will compute alphas for the three revised math attitude scales (*motivation*, *competence*, and *pleasure*) that items 1 to 14 were intended to assess.

Assumptions for Measures of Reliability

When two or more items or assessments are viewed as measuring the same underlying variable (construct), reliability can be assessed; for example, if on a questionnaire four items were intended to measure stress, you could assess the reliability of the resulting data. Reliability is used to indicate the extent to which scores are consistent with one another and the extent to which the data are free from measurement error. It is assumed that each item or score is composed of a true score measuring the underlying construct, plus error; there is almost always some error in the measurement. Therefore, one assumption is that the measures or items are related systematically to one another in a linear manner because they are believed to be measures of the same construct.

Measurement Validity

In this chapter, we discuss one method to provide evidence supporting measurement validity. Measurement validity is concerned with establishing evidence for the use of a measure or instrument in a particular setting with a specific population for a given purpose. We use the term measurement validity; others might use terms such as *test validity*, *score validity*, or just *validity*. We use the modifier *measurement* to point out that <u>the scores only provide evidence for validity; it is inappropriate to say that a test or measure is "valid" or "invalid."</u> Thus, when we address the issue of measurement validity with respect to a particular test, we are addressing the issue of the evidence for the validity of the scores on that test for a particular purpose, and not the validity of the test or instrument in general.

Reliability or consistency is a necessary prerequisite for measurement validity. However, <u>an instrument may</u> produce consistent data <u>(provide evidence for reliability), but the data may not be valid</u> because they are not an accurate measure of the intended concept. In research articles, there is usually more evidence for the reliability of the instrument than for the validity of the instrument because evidence for validity is more difficult to obtain. To establish validity, one ideally needs a "gold standard" or "criterion" related to the particular purpose of the measure. To obtain such a criterion is often not an easy matter or sometimes even possible, so other types of evidence to support the validity of a measure are necessary. Increasingly, validity has been conceptualized as a unitary concept; however, many types of evidence should be gathered to help assess validity for a given set of data.

Several broad types of evidence to support validity are discussed in research methods textbooks. No one type of evidence alone is sufficient. Validation should integrate all the pertinent evidence from as many of the types of evidence as possible. Preferably, validation should include some evidence in addition to content evidence, which is probably the most common and easiest to obtain. **Content evidence** refers to whether the content that makes up the instrument is a reasonable representation of the concept that one is attempting to measure. Does the instrument accurately represent the major aspects of the concept and not include material that is irrelevant to it? There is no statistic that demonstrates evidence based on the content of the measure. Because this type of evidence depends on the logical, but subjective, agreement of a few experts, we consider it necessary but not sufficient evidence. The experts review the measure for clarity and fit with the construct to be measured. The 14 math attitude questions presented in Chapter 1 and used in Problems 4.3 and 4.4 were developed and refined using experts to provide evidence for content validity.

Evidence Based on Internal Structure

An example of internal structure evidence will be demonstrated in this chapter. This is a type of evidence for validity, not a separate indicator of validity. Evidence from several types of analysis, including factor analysis, can be useful here. Most questionnaires have an overall construct to be measured; in this chapter, the example is student self-perceptions regarding mathematics. Many times, the overall construct will have subconstructs; multiple areas that measure aspects of the overall construct. In this example, the subconstructs were the perceptions of math competence, math motivation, and pleasure doing math. **Factor analysis,** demonstrated in Problem 4.3, can provide evidence for validity based on internal structure when a construct is complex and several aspects (or factors) of it are measured. If the clustering of items supports the theory-based grouping of items, factorial evidence of validity is provided. Therefore, from this example, a factor analysis would help us identify the degree to which the data supported the subconstructs by indicating if the respondents' answers to the questions for competence were relatively highly correlated with each other and less highly correlated with the questions for motivation and for pleasure.

- Retrieve **hsbdataNEW** from your data file. See Appendix A for how to retrieve the dataset if you need to.

Problem 4.1: Cohen's Kappa to Assess Reliability With Nominal Data

When we have two nominal variables with the *same* possible values (usually two raters' observations or scores using the same codes) you can compute Cohen's kappa (also referred to as kappa) to check the interrater reliability or agreement between the codes. Kappa has a few assumptions about the underlying nature of the data: (a) participants are independent of each other, (b) the raters, reporters, or observers providing the data do so independently of one another, and (c) the rating categories are mutually exclusive and exhaustive. Data typically are nominal, but they can be ordinal; however, kappa treats them as unordered. If the data are normally distributed, other statistics are preferable. Imagine that the ethnicity variable was based on school records. Then, a new variable was obtained by asking students to self-report their ethnicity. Ethnicity does involve some subjectivity, especially if only one ethnicity can be used to represent each person when they may identify with more than one ethnicity. The question is, how reliable is the interobserver classification data for ethnicity?

4.1. What is the interrater reliability coefficient for the *ethnicity* codes (based on school records) and *ethnicity reported by the student*? Is there consistency between the two measures of ethnicity?

To compute kappa:

- Click on **Analyze → Descriptive Statistics → Crosstabs. . .** to get Figure 4.1.
- Click on **Reset** to clear any previous entries.
- Move *ethnicity (ethnic)* to the **Rows** box and *ethnicity reported by students (ethnic 2)* to the **Columns** box. (See Figure 4.1.)

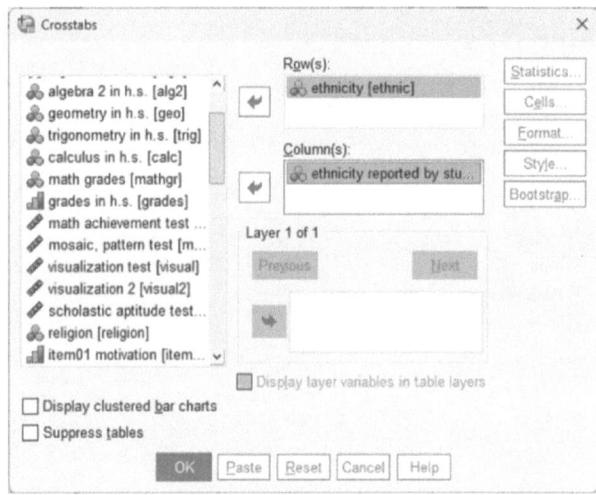

Figure 4.1 Crosstabs.

- Click on **Statistics** in the upper right corner of Figure 4.1 to get Figure 4.2.
- Click on **Kappa** in the **Statistics** dialog box in Figure 4.2.

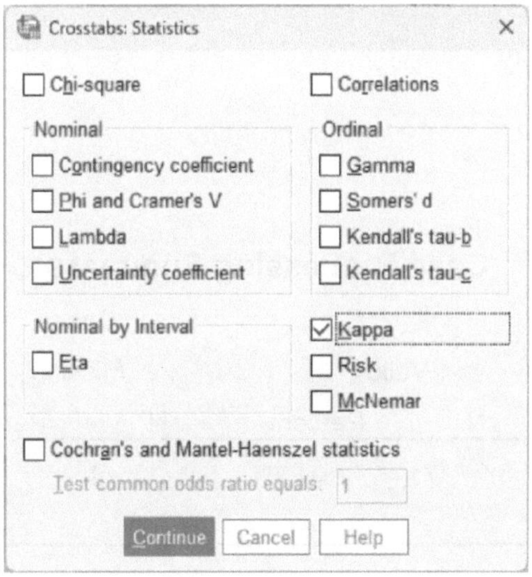

Figure 4.2 Crosstabs statistics.

- Click on **Continue** to go back to the Crosstabs window (see Figure 4.1).
- Then click on **Cells** in Figure 4.1 to get Figure 4.3 and request the **Observed** under **Counts** and **Total** under **Percentages**. Leave the rest of the window as is.
- Click on **Continue** and then **OK**. Compare your syntax and output to Output 4.1.

Output 4.1: Cohen's Kappa With Nominal Data

```
CROSSTABS
  /TABLES=ethnic BY ethnic2
  /FORMAT= AVALUE TABLES
  /STATISTICS=KAPPA
  /CELLS= COUNT TOTAL
  /COUNT ROUND CELL.
```

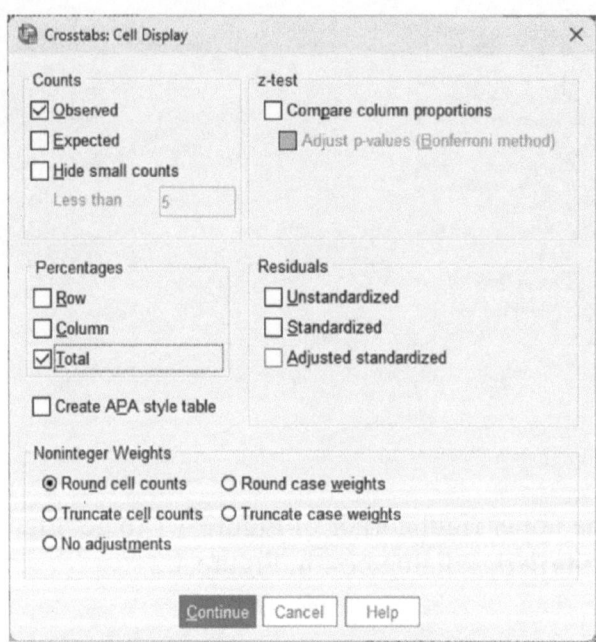

Figure 4.3 Crosstabs: Cell display.

Crosstabs

Case Processing Summary

	Cases					
	Valid		Missing		Total	
	N	Percent	N	Percent	N	Percent
ethnicity * ethnicity reported by student	71	94.7%	4	5.3%	75	100.0%

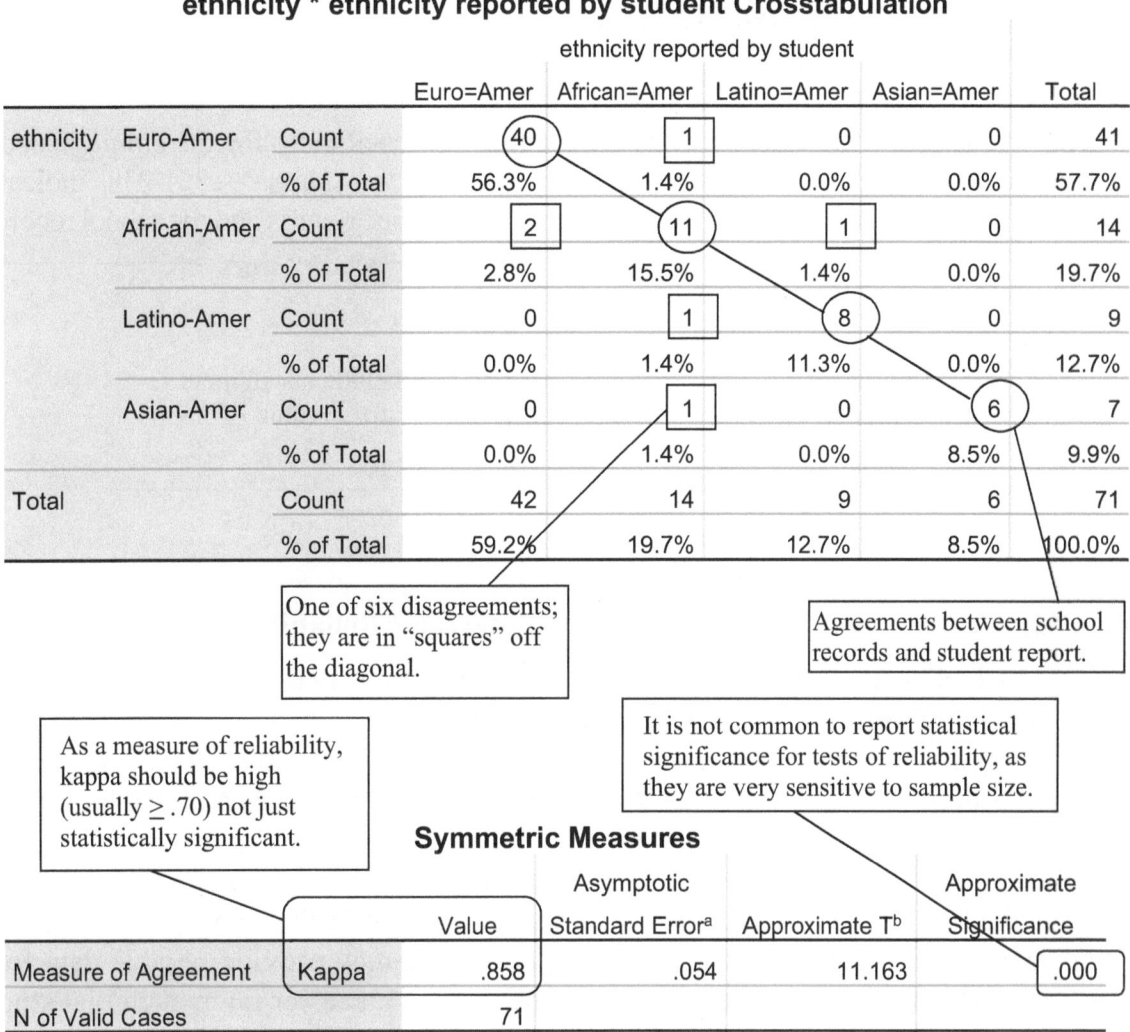

ethnicity * ethnicity reported by student Crosstabulation

			ethnicity reported by student				Total
			Euro=Amer	African=Amer	Latino=Amer	Asian=Amer	
ethnicity	Euro-Amer	Count	40	1	0	0	41
		% of Total	56.3%	1.4%	0.0%	0.0%	57.7%
	African-Amer	Count	2	11	1	0	14
		% of Total	2.8%	15.5%	1.4%	0.0%	19.7%
	Latino-Amer	Count	0	1	8	0	9
		% of Total	0.0%	1.4%	11.3%	0.0%	12.7%
	Asian-Amer	Count	0	1	0	6	7
		% of Total	0.0%	1.4%	0.0%	8.5%	9.9%
Total		Count	42	14	9	6	71
		% of Total	59.2%	19.7%	12.7%	8.5%	100.0%

One of six disagreements; they are in "squares" off the diagonal.

Agreements between school records and student report.

As a measure of reliability, kappa should be high (usually ≥ .70) not just statistically significant.

It is not common to report statistical significance for tests of reliability, as they are very sensitive to sample size.

Symmetric Measures

		Value	Asymptotic Standard Error[a]	Approximate T[b]	Approximate Significance
Measure of Agreement	Kappa	.858	.054	11.163	.000
N of Valid Cases		71			

a. Not assuming the null hypothesis.

b. Using the asymptotic standard error assuming the null hypothesis.

Interpretation of Output 4.1

The **Case Processing Summary** table shows that 71 students have data on both variables. The **Cross-tabulation** table shows that six students' codes disagree with the school records, as indicated by boxes off the diagonal; this table also shows the cases where the school records and the student self-reports are in agreement (they are on the diagonal and circled). There are 65 (40 + 11 + 8 + 6) students with such agreement or consistency.

The **Symmetric Measures** table shows that kappa is .86. This indicates good interobserver reliability because such measures should be high (≥ .70) and positive. Statistical significance is not relevant for reliability measures.

Example of How to Write About Problem 4.1

Method

Kappa was used to investigate the interrater reliability coefficient for the *ethnicity* codes (based on school records) and *ethnicity reported by the student* (kappa = .86). This indicates that there is strong evidence for reliability between the students' reports and the school records.

Remember to report the exact value of kappa.

Be sure to include a statement regarding what the value of kappa means.

Problem 4.2: Correlation and Paired *t* to Assess Interrater Reliability

As mentioned earlier, there are several ways to assess interrater reliability when one has normally distributed (scale) data. One way is to use a paired *t* test. We could demonstrate interrater reliability for the *visualization test* scores using correlation. However, the paired computing test in SPSS may be a better way to go because it produces and displays not only the reliability <u>correlation</u> but also the comparison of the <u>means</u> for the two raters. Thus, the correlation will tell us whether the test scores were strongly associated. In addition, the *t* test will tell us whether, on average, scores of the second rater were the same (versus systematically higher or lower) than the first rater. Thus, two raters may provide reliable data for the same construct (high positive correlation), but one may be an easier rater, such that students seem to perform at a higher level. The paired *t* enables one to determine this, providing more information about the tests. Another advantage of the paired *t* is that several such reliability checks can be made with one SPSS run with a relatively compact and easily interpretable output. Here we only check the reliability of two variables, *visualization test* and *mosaic pattern test*. Each of these tests had somewhat subjective scoring so were observed by two researchers in order to check interrater reliability.

4.2. What is the interrater reliability of the *visualization test* and *mosaic pattern test* scores? Do average *visualization test* and *mosaic pattern test* scores by the second rater differ from average visualization test and mosaic pattern test scores by the first rater?

To compute reliability and mean differences with the paired *t* test program:

- Select **Analyze → Compare Means → Paired Samples T Test. . .**
- Click on both *visualization test* and *visualization 2* and move them to the **Paired Variables:** box.
- Next, click on *mosaic pattern test* and *mosaic pattern test 2* and move them to the **Paired Variables** box (see Figure 4.4).

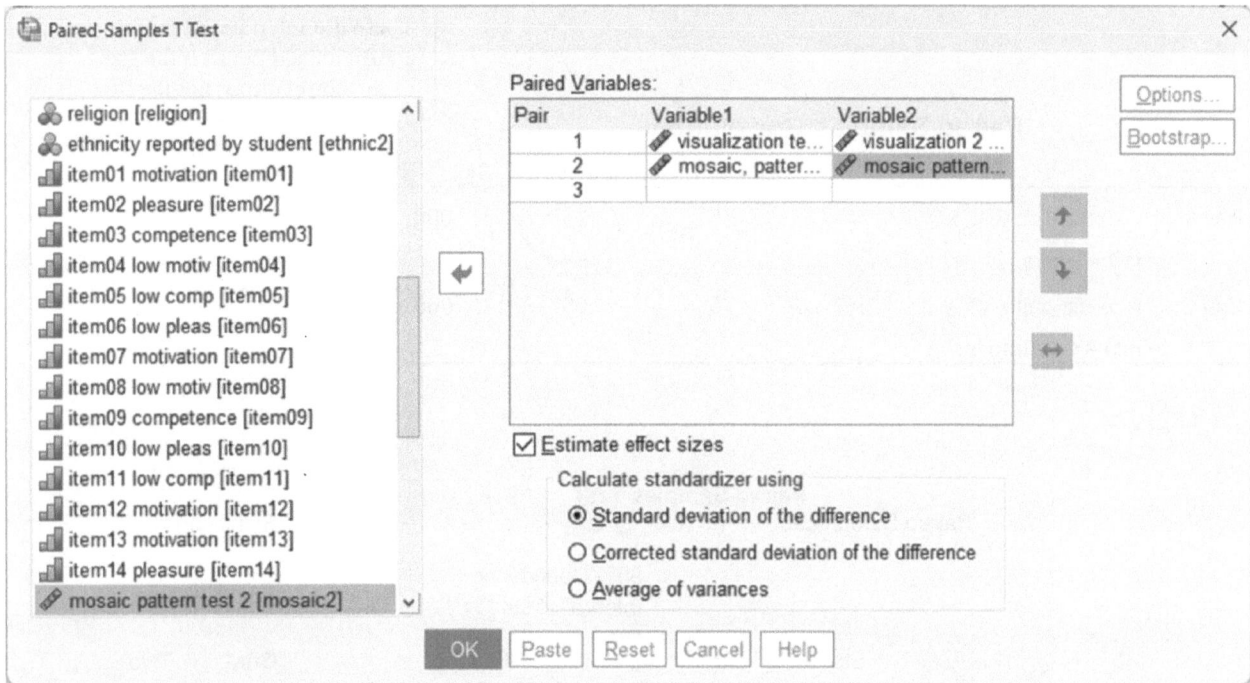

Figure 4.4 Paired-samples t test.

- Click on **OK**.

Compare your output to Output 4.2.

Output 4.2: Test–Retest Reliability for Visualization Scores

```
T-TEST PAIRS=visual mosaic WITH visual2 mosaic2 (PAIRED)
  /ES DISPLAY(TRUE) STANDARDIZER(SD)
  /CRITERIA=CI(.9500)
  /MISSING=ANALYSIS.
```

T-Test

Paired Samples Statistics

		Mean	N	Std. Deviation	Std. Error Mean
Pair 1	visualization test	5.2433	75	3.91203	.45172
	visualization 2	5.1067	75	3.77518	.43592
Pair 2	mosaic, pattern test	27.413	75	9.5738	1.1055
	mosaic pattern test 2	27.4800	75	9.34816	1.07943

Notice that the means for *visualization test* and *visualization 2* and also for *mosaic pattern test* and *mosaic pattern test 2* are very similar.

These are interrater reliability coefficients. Focus on the <u>size</u> of the correlations, not the Sig.

Paired Samples Correlations

		N	Correlation	Sig.
Pair 1	visualization test & visualization 2	75	.938	.000
Pair 2	mosaic, pattern test & mosaic pattern test 2	75	.957	.000

Paired Samples Test

		Paired Differences							Significance	
					95% Confidence Interval of the Difference					
		Mean	Std. Deviation	Std. Error Mean	Lower	Upper	t	df	One-Sided p	Two-Sided p
Pair 1	visualization test-visualization 2	.13667	1.36331	.15742	-.17700	.45033	.868	74	.194	.388
Pair 2	mosaic, pattern test - mosaic pattern test 2	-.06667	2.77334	.32024	-.70476	.57142	-.208	74	.418	.836

			Standardizer[a]	Point Estimate	95% Confidence Interval	
					Lower	Upper
Pair 1	visualization test - visualization 2	-Cohen's d	1.36331	.100	-.127	.327
		Hedges' correction	1.37732	.099	-.126	.323
Pair 2	mosaic, pattern test - mosaic pattern test 2	-Cohen's d	2.7733	-.024	-.250	.202
		Hedges' correction	2.8019	-.024	-.248	.200

a. The denominator used in estimating the effect sizes.
Cohen's d uses the sample standard deviation of the mean difference.
Hedges' correction uses the sample standard deviation of the mean difference, plus a correction factor.

Interpretation of Output 4.2

The first table, **Paired Samples Statistics**, shows the **Means** for the two raters of the students' *visualization test* and of their *mosaic pattern test*. These means will be compared in the third table. In addition, the *N*s, *SD*s, and standard errors are shown.

The second table shows the **Paired Samples Correlations**, which will be used to assess the interrater reliabilities of the *visualization test* scores and the *mosaic pattern test* scores. Note that *r*s are .94 and .96, which are very high positive correlations and provide good support for reliability. These correlations indicate that students who were scored high by the first observer on either test were very likely to be scored high by the second observer, and students who were scored low were very likely to be scored low. It indicates that the visualization and mosaic tests are systematically measuring primarily the same thing no matter which observer (rater) is scoring them. However, it cannot indicate *what construct or skill* the raters' or observers' scores are measuring. This would be the goal of *validity* measurement.

The **Paired Samples Test** table shows that the means of the two raters of the *visualization test* and the two raters of the *mosaic pattern test* provide very similar average scores; the means are not significantly different ($p = .39$ and .84). It is usually desirable for the means of two assessments of the same variable to be similar, indicating that one rater was not a harder or easier scorer (grader) than the other.

Example of How to Write About Problem 4.2

Method

The interrater reliability coefficient for both the visualization test data, $r(73) = .94$, and the mosaic pattern test data, $r(73) = .96$, indicated strong evidence for reliability.

You should also include information from the literature about the scales (e.g., past reliability and validity) in other studies.

Make note of the spacing: per APA guidelines. You do not include a space between the *r* and the *df*. You do include a space on both sides of equal sign (=).

Problem 4.3: Exploratory Factor Analysis to Assess Evidence for Validity

In Problem 4.3, we perform a principal axis factor analysis on the math attitude variables. Exploratory factor analysis (EFA) is appropriate when one has the belief that there are latent constructs underlying the variables; usually these constructs are measured by groups of items. In this example, we have beliefs about the constructs underlying the math attitude questions; we believe that there are three constructs: *motivation scale*, *competence scale*, and *pleasure scale*. Now, we want to see if the items that were written to measure each of these constructs actually do "hang together." That is, we wish to determine empirically whether participants' responses to the motivation questions are more similar to each other than to their responses to the competence items, and so on. Conducting factor analysis can assist us in validating the data. If the data do fit reasonably well into the three constructs that we believe exist, then this gives us internal structure evidence to support the validity of the math attitude measures in this sample. We "allow" the factor analysis to find factors that best fit the data, even if this deviates from our original predictions.

Conditions and Assumptions for Factor Analysis

There are two main conditions necessary for factor analysis. The first is that there need to be relationships among the variables or items. Further, the larger the sample size, especially in relation to the number of variables, the more reliable the resulting factors. Principal axis factor analysis, used in Problem 4.3, should never be used if the number of items (variables) is greater than the number of participants. The methods of extracting factors that are used in this book do not make strong distributional assumptions; normality is important only to the extent that skewness or outliers affect the observed correlations. Because factor analysis is based on correlations, independent sampling is required, and the variables should be related to each other (in pairs) in a linear fashion. Each of the variables should be correlated at a moderate level with some of the other variables. Factor analysis would not be a sensible thing to do if the correlations all hover around zero. Bartlett's test of sphericity addresses this assumption. However, if correlations are all too high, this may cause problems with obtaining a mathematical solution; this is tested with the Determinant. See the Interpretation of Output 4.3 for more testing assumptions.

The question for Problem 4.3 could be stated as follows:

4.3. Do the results provide evidence to support the validity of our conceptualization of three factors (*competence scale*, *motivation scale*, and *pleasure scale*) underlying the math attitude questions?

To answer this question, we will conduct a factor analysis using the principal axis factoring method and specify the number of factors to be three (because our conceptualization is that there are three math attitude scales or factors: *motivation scale*, *competence scale*, and *pleasure scale*).

- **Analyze → Dimension Reduction → Factor . . .** to get Figure 4.5.
- Next, select the variables *item01 motivation* through *item14 pleasure*. Do not include *item04 reversed* or any of the other reversed items because we are including the unreversed versions of those same items.

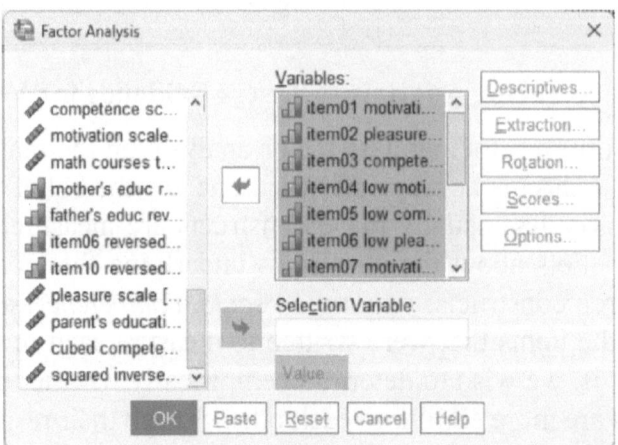

Figure 4.5 Factor analysis.

- Now click on **Descriptives . . .** to produce Figure 4.6.
- Under **Statistics** click **Univariate descriptives** and *unclick* **Initial solution** if it is checked.
- Then click on the following: **Coefficients**, **Determinant**, and **KMO and Bartlett's test of sphericity** (under **Correlation Matrix**).
- Click on **Continue** to return to Figure 4.5.

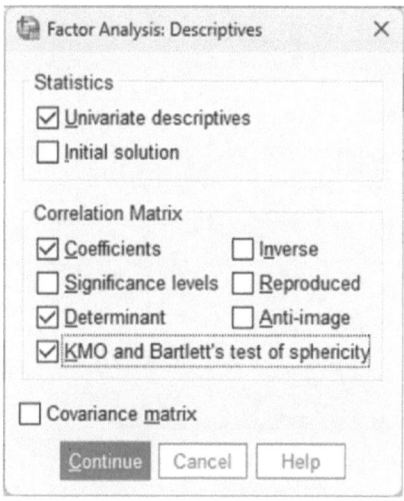

Figure 4.6 Factor analysis: Descriptives.

- Next, click on **Extraction . . .** This will give you Figure 4.7.
- Select **Principal axis factoring** from the **Method** pull-down menu.
- *Unclick* **Unrotated factor solution** (under **Display**).
- Click on **Fixed number of factors** under **Extract**, and type **3** in the box. This setting instructs the computer to extract three factors, which we are hoping will correspond to the three predicted math perception constructs.
- Click on **Continue** to return to Figure 4.5.

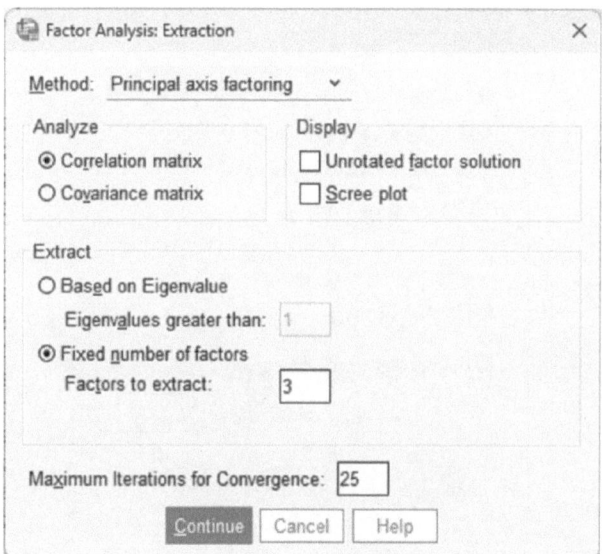

Figure 4.7 Extraction method to produce principal axis factoring.

- Now click on **Rotation . . .** in Figure 4.5, which will give you Figure 4.8.
- Click on **Varimax**, then make sure **Rotated solution** is also checked. Varimax rotation creates a solution in which the factors are orthogonal (uncorrelated with one another), which can make results easier to interpret.
- Click on **Continue** to go back to Figure 4.5.

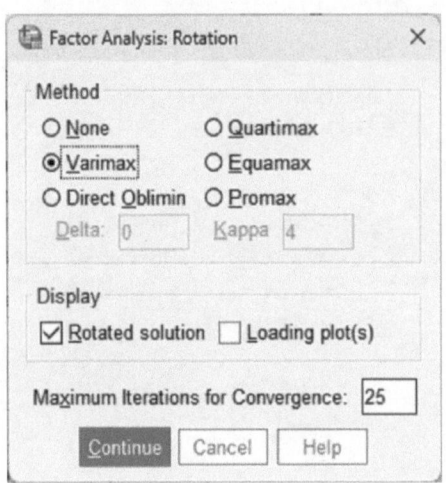

Figure 4.8 Factor analysis: Rotation.

- Next, click on **Options . . .** which will give you Figure 4.9.
- Click on **Sorted by size**.
- Click on **Suppress small coefficients . . . Absolute value below** and type **.3** (point 3) in the box (see Figure 4.9). Suppressing small factor loadings makes the output easier to read.
- Click on **Continue** then **OK**. Compare Output 4.3 with your output and syntax.

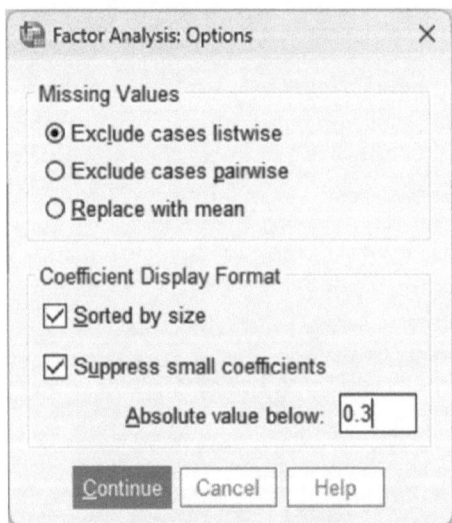

Figure 4.9 Factor analysis: Options.

Output 4.3: Factor Analysis for Math Attitude Questions

```
FACTOR
  /VARIABLES item01 item02 item03 item04 item05 item06 item07 item08
item09 item10 item11 item12 item13 item14
  /MISSING LISTWISE
  /ANALYSIS item01 item02 item03 item04 item05 item06 item07 item08
item09 item10 item11 item12 item13 item14
  /PRINT UNIVARIATE CORRELATION DET KMO ROTATION
  /FORMAT SORT BLANK(0.3)
  /CRITERIA FACTORS(3) ITERATE(25)
  /EXTRACTION PAF
  /CRITERIA ITERATE(25)
  /ROTATION VARIMAX
  /METHOD=CORRELATION.
```

Factor Analysis

Descriptive Statistics

	Mean	Std. Deviation	Analysis N
item01 motivation	2.99	.918	71
item02 pleasure	3.58	.822	71
item03 competence	2.82	.915	71
item04 low motiv	2.21	.909	71
item05 low comp	1.61	.948	71
item06 low pleas	2.44	.996	71
item07 motivation	2.77	1.072	71
item08 low motiv	1.96	.917	71
item09 competence	3.32	.770	71
item10 low pleas	1.41	.748	71
item11 low comp	1.38	.763	71
item12 motivation	2.99	.837	71
item13 motivation	2.68	.807	71
item14 pleasure	2.86	.723	71

Note that Factor Analysis and most complex analyses only use participants with no missing data on any of the variables, so in this case N = 71.

Pairs of items with relatively high (+ or -) correlations (e.g., > .40) will probably have high loadings from the same factor.

This piece of the matrix indicates how each question is associated (correlated) with each of the other questions.

Correlation Matrix[a]

		item01 motivation	item02 pleasure	item03 competence	item04 low motiv	item05 low comp	item06 low pleas	item07 motivation	item08 low motiv	item09 competence
Correlation	item01 motivation	1.000	.484	.626	-.305	-.745	-.165	.461	-.340	.209
	item02 pleasure	.484	1.000	.389	-.166	-.547	-.312	.361	-.176	.219
	item03 competence	.626	.389	1.000	-.348	-.743	-.209	.423	-.248	.328
	item04 low motiv	-.305	-.166	-.348	1.000	.363	.323	-.596	.576	-.120
	item05 low comp	-.745	-.547	-.743	.363	1.000	.260	-.538	.276	-.351
	item06 low pleas	-.165	-.312	-.209	.323	.260	1.000	-.268	.192	-.131
	item07 motivation	.461		.423	-.596	-.538	-.268	1.000	-.606	.228
	item08 low motiv	-.340	-.176	-.248	.576	.276	.192	-.606	1.000	-.243
	item09 competence	.209	.219	.328	-.120	-.351	-.131	.228	-.243	1.000
	item10 low pleas	.071	-.389	.027	.102	.130	.217	-.169	.067	-.109
	item11 low comp	-.441	-.401	-.513	.398	.605	.418	-.331	.370	-.407
	item12 motivation	.186	.116	.165	-.391	-.187	-.044	.347	-.392	.406
	item13 motivation	.187	.028	.170	-.334	-.169	.001	.361	-.308	.286
	item14 pleasure	.040	.475	.068	-.063	-.166	-.469	.180	-.117	-.020

a. Determinant = .001

Should be greater than .0001. If very close to zero, collinearity is too high. If zero, no solution is possible.

Low correlations (e.g., < .20) usually will not have high loadings from the same factor.

Tests of assumptions.

Should be greater than .70 indicating sufficient items for each factor.

KMO and Bartlett's Test

Kaiser-Meyer-Olkin Measure of Sampling Adequacy.		.770
Bartlett's Test of Sphericity	Approx. Chi-Square	433.486
	df	91
	Sig.	.000

Should be significant (less than .05), indicating that the correlation matrix is significantly different from an identity matrix, in which correlations between variables are all zero.

Interpretation of Output 4.3

The SPSS factor analysis program generates a variety of tables depending on which options you have chosen. Our first table includes **Descriptive Statistics** for each variable and the **Analyses N**, which in this case is 71 because several items have one or more participants missing. It is especially important to check the Analysis N when you have a small sample, scattered missing data, or one variable with lots of missing data. In the latter case, it may be wise to run the analysis without that variable. The second table in Output 4.3 is a **correlation matrix** showing how each of the 14 items is associated with each of the other 13. Note that some of the correlations are high (e.g., + or −.60 or greater) and some are low (i.e., nears zero). Relatively high correlations indicate that two items are associated and will probably be grouped together by the factor analysis.

Next, several assumptions are tested. The **determinant** (located under the correlation matrix) should be more than .0001 or collinearity is too high. Here, it is .001 so this assumption is met. The **Kaiser-Meyer-Olkin (KMO)** measure should be greater than .70 and is inadequate if less than .50. The KMO test tells us whether or not enough items are predicted by each factor. The **Bartlett** test should be significant (i.e., a significance value of less than .05); this means that the variables are correlated highly enough to provide a reasonable basis for factor analysis.

Factor Matrix[a]

Rotated Factor Matrix[a]

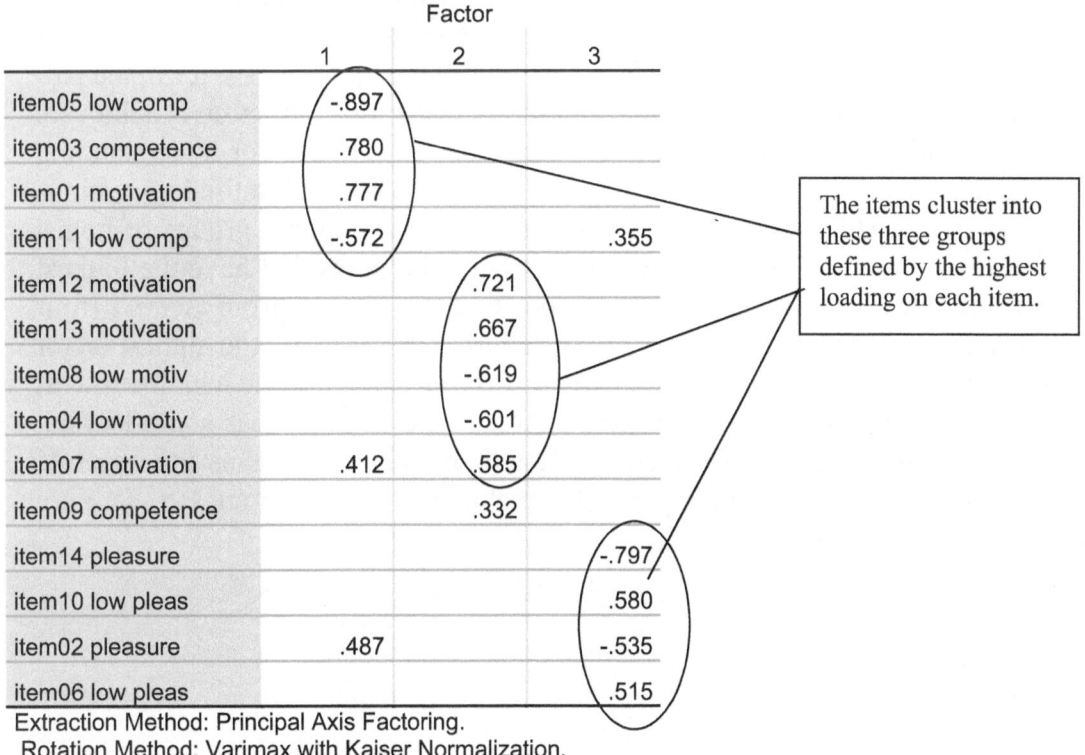

	Factor 1	Factor 2	Factor 3
item05 low comp	-.897		
item03 competence	.780		
item01 motivation	.777		
item11 low comp	-.572		.355
item12 motivation		.721	
item13 motivation		.667	
item08 low motiv		-.619	
item04 low motiv		-.601	
item07 motivation	.412	.585	
item09 competence		.332	
item14 pleasure			-.797
item10 low pleas			.580
item02 pleasure	.487		-.535
item06 low pleas			.515

The items cluster into these three groups defined by the highest loading on each item.

Extraction Method: Principal Axis Factoring.
Rotation Method: Varimax with Kaiser Normalization.
a. Rotation converged in 5 iterations.

Total Variance Explained

This table shows the percentage of variance explained by each of the three factors and the cumulative percentage, which is 51%.

Factor	Total	% of Variance	Cumulative %
1	3.017	21.549	21.549
2	2.327	16.621	38.171
3	1.784	12.746	50.917

Rotation Sums of Squared Loadings

Half of the variance is accounted for by the first three factors.

Extraction Method: Principal Axis Factoring.

Factor Transformation Matrix

We will ignore this; it was used to convert the initial factor matrix into the rotated factor matrix.

Factor	1	2	3
1	.747	.552	-.370
2	-.162	.692	.704
3	.645	-.466	.606

Extraction Method: Principal Axis Factoring.
Rotation Method: Varimax with Kaiser Normalization.

Interpretation of Output 4.3 continued

We rotate the factors so that they are easier to interpret. Rotation makes it so that, as much as possible, different items are explained or predicted by different underlying factors, and each factor explains more than one item. This is a condition called simple structure. Although this is the *goal* of rotation, in reality, this is not always achieved. One thing to look for in the **Rotated Factor Matrix** of factor loadings is the extent to which simple structure is achieved. The Rotated Factor Matrix table, which contains these loadings, is key for understanding the results of the analysis. Note that the analysis has sorted the 14 math attitude questions (*item01 motivation* to *item14 pleasure*) into three somewhat overlapping groups of items, as shown by the circled items. The items are sorted so that the items that have the highest loading (not considering whether the correlation is positive or negative) from factor 1 (four items in this analysis) are listed first, and they are sorted from the one with the highest factor weight or loading (i.e., *item05 low comp*, with a loading of −.897) to the one with the lowest loading **from that first factor** (*item11 low comp*). Actually, every item has some loading from every factor, but we requested for loadings less than |.30| (|.30| means the absolute value, or value without considering the sign) to be excluded from the output, so there are blanks where low loadings exist.

Next, the six items that have their highest loading from factor 2 are listed from highest loading (*item12 motivation*) to lowest (*item9 competence*). Finally, the four items on which factor 3 loads most highly are listed in order. Loadings resulting from an orthogonal rotation are correlation coefficients between each item and the factor, so they range from −1.0 through 0 to +1.0. A negative loading just means that the question needs to be interpreted in the opposite direction from the way it is written for that factor (e.g., *item05* "I am a little slow catching on to new topics in math" has a negative loading from the competence factor, which indicates that the people scoring **higher** on this item are **lower** in competence). Usually, factor loadings lower than |.30| are considered low, which is why we suppressed loadings less than |.30|. On the other hand, loadings of |.40| or greater are typically considered high. This is just a guideline, however, and one could set the criterion for "high" loadings as low as .30 or as high as .50. Setting the criterion lower than .30 or higher than .50 would be very unusual.

The investigator should examine the content of the items that have high loadings from each factor to see if they fit together conceptually and can be named. For example, items 5, 3, and 11 were intended to reflect perceptions of *competence* at math, so the fact that they all have strong loadings on the same factor provides some support for their being conceptualized as pertaining to the same construct. On the other hand, *item01 motivation* was intended to measure motivation for doing math, but it loaded highly on this same competence factor. In retrospect, one can see why this item could also be interpreted as competence. The item reads, "I practice math skills until I can do them well." Unless one felt one **could do** math problems well, this would not be true. So it is not surprising that *item01* loaded on the *competence* factor. Likewise, *item02 pleasure*, "I feel happy after solving a hard problem," although intended to measure pleasure at doing math (and having its strongest loading there), the question might also reflect competence at doing math, in that, again, one could not endorse this item unless one had solved hard problems, which one could only do if one were good at math. On the other hand, *item09 competence*, which was originally conceptualized as a competence item, had no really strong loadings so we will not use it as part of the revised motivation scale in Problem 4.4.

Every item has a weight or loading from every factor, but in a "clean" factor analysis almost all of the loadings that are **not** in the circles that we have drawn on the **Rotated Factor Matrix** will be low (blank or less than |.40|). The fact that both Factors 1 and 3 load highly on *item02 pleasure*, and the fact that Factors 1 and 2 both load highly on *item07 motivation* is **common but undesirable**, in that one wants only one factor to predict each item. Thus, with modest exceptions, this factor analysis provides internal structure evidence to support the validity of the three math attitude concepts.

The last table we consider is the **Total Variance Explained** table. It shows that the percentage of variance explained by the three factors after rotation is 21.5%, 16.6%, and 12.7% with a cumulative total of 50.9% explained by the three factors.

Always include the specific name of the analysis that was conducted.

Be sure to present the variance that each subconstruct accounts for.

How to Write About Problem 4.3

Method

Principal axis factor analysis with varimax rotation was conducted to assess the underlying structure for the 14 items of the Math Motivation Questionnaire. Three factors were requested, based on the fact that the items were designed to index three constructs: motivation, competence, and pleasure. After rotation, the first factor accounted for 21.5% of the variance, the second factor accounted for 16.6%, and the third factor accounted for 12.7%. Table 4.1 displays the items and factor loadings for the rotated factors, with loadings less than .40 omitted to improve clarity.

The first factor, which seems to index competence, had strong loadings on the four items. Two of the items indexed low competence and had negative loadings. The second factor, which seemed to index motivation, had high loadings on the next five items in Table 4.1. "I prefer to figure out the problem without help" had its highest loading from the second factor but had a cross-loading over .4 on the competence factor. The third factor, which seemed to index low pleasure from math, loaded highly on the last four items in the table. "I feel happy after solving a hard problem" had its highest loading from the pleasure factor but also had a strong loading from the competence factor. Thus, the result provides some support for validity; namely, that there are three concepts (competence, motivation, and pleasure) measured by the 14 items.

Table 4.1

Factor Loadings for the Rotated Factors

Always include the rotated factor loadings in a table.

Item	Factor Loadings		
	1	*2*	*3*
Slow catching on to new topics	−.90		
Solve math problems quickly	.78		
Practice math until do well	.78		
Have difficulties doing math	−.57		
Try to complete math even if takes long		.72	
Explore all possible solutions		.67	
Do not keep at it long if problem challenging		−.62	
Give up easily instead of persisting		−.60	
Prefer to figure out problems without help	.41	.59	
Really enjoy working math problems			−.80
Smile only a little when solving math problem			.58
Feel happy after solving hard problem	.49		−.54
Do not get much pleasure out of math			.52
% of variance	21.55	16.62	12.75

Note. Loadings < .40 are omitted.

Problem 4.4: Cronbach's Alpha to Assess Internal Consistency Reliability

The 14 items used in the factor analysis seem to fit best into three groups that could be labeled, as we hypothesized, competence, motivation, and pleasure. Based on the factor analysis, competence is composed of four items that were rated on Likert scales, from very atypical (1) to very typical (4). Do the scores for these items go together (interrelate) well enough to add them together for future use as a composite or summated variable labeled *revised competence*? Note that this revised scale is somewhat different from the one in Chapter 1 and used later in this book because it does not include *item 09 competence* but does include *item 01 motivation* which was intended to be a motivation item.

4.4a. What is the internal consistency reliability of the math attitude scale data that we have labeled *revised competence*?

Note that you <u>do not use the computed scale score</u>. Instead, use the individual items to create the scale temporarily. Now let's do reliability analysis for the revised competence scale.

- Click on **Analyze → Scale → Reliability Analysis**. You should get a dialog box like Figure 4.10.
- Now move the variables *item01 motivation, item03 competence, item05 reversed, and item11 reversed* (the competence questions based on the factor analysis) to the **Items** box. Be sure to use items 05 and 11 reversed (<u>not</u> *item05 low comp* and *item11 low comp*) because a high rating on the original (unreversed) items indicates low motivation. The alpha will be based on the average correlation among each pair of items, so they all need to be scored so that higher scores index the same thing (e.g., higher levels of competence) so you won't be averaging positive and negative correlations.
- Type **Alpha for the Revised Competence Scale** in the **Scale label** box. Be sure the **Model** is **Alpha** (refer to Figure 4.10).

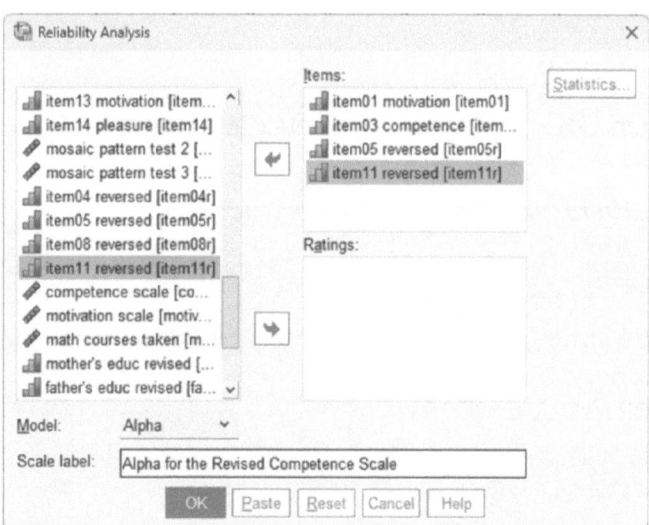

Figure 4.10 Reliability analysis.

- Click on **Statistics** in the **Reliability Analysis** dialog box and you will see something similar to Figure 4.11.
- Check the following items: **Item** and **Scale if item deleted** (all under Descriptives for) and **Correlations** (under Inter-Item).

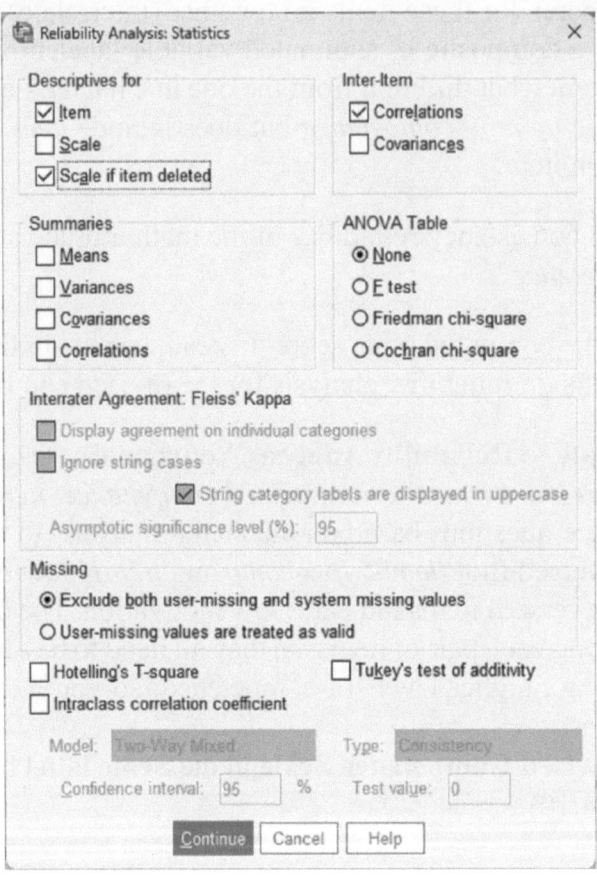

Figure 4.11 Reliability analysis: Statistics.

- Click on **Continue** then **OK**. Compare your syntax and output to Output 4.4a.

Output 4.4a: Cronbach's Alpha for the Revised Competence Scale

```
RELIABILITY
  /VARIABLES=item01 item03 item05r item11r
  /SCALE('Alpha for the Revised Competence Scale') ALL
  /MODEL=ALPHA
  /STATISTICS=DESCRIPTIVE CORR
  /SUMMARY=TOTAL.
```

Reliability Scale: Alpha for the Revised Competence Scale

Case Processing Summary

		N	%
Cases	Valid	73	97.3
	Excluded[a]	2	2.7
	Total	75	100.0

a. Listwise deletion based on all variables in the procedure.

Reliability Statistics

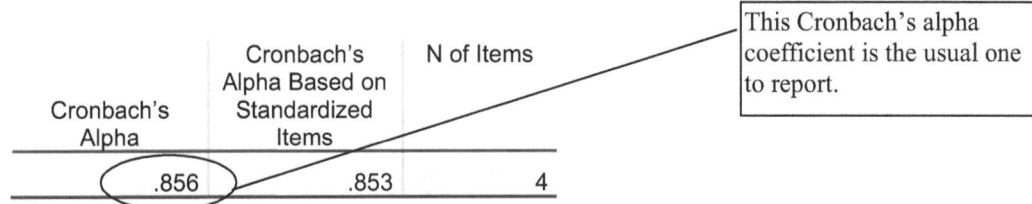

Cronbach's Alpha	Cronbach's Alpha Based on Standardized Items	N of Items
.856	.853	4

This Cronbach's alpha coefficient is the usual one to report.

Item Statistics

	Mean	Std. Deviation	N
item01 motivation	2.96	.934	73
item03 competence	2.82	.903	73
item05 reversed	3.37	.979	73
item11 reversed	3.63	.755	73

Descriptive statistics for each variable. Note 73 of the participants have data on all 4 items.

Interpretation of Output 4.4a

The first table shows the number of **Valid** cases, that is, those with no missing data on the selected variables. The second table lists the **Cronbach's Alpha** and an **Alpha Based on Standardized Items**. In general, you will use the unstandardized alpha unless the items in the scale have quite different means and standard deviations. For example, if one were to compute an alpha for *math achievement test* ($M = 12.6$, $SD = 6.7$), *grades in h.s.* ($M = 5.7$, $SD = 1.6$), and *visualization test* ($M = 5.2$, $SD = 3.9$), we would use the standardized alpha.

As with other reliability coefficients, alpha should be above .70; however, it is common to see journal articles where one or more scales have somewhat lower alphas (e.g., in the .60–.69 range), especially if there is only a handful of items in the scale. A very high alpha (e.g., greater than .90) probably means that the items are repetitious or that you have more items in the scale than are really necessary for an internally reliable measure of the concept. In this example, the alpha for the *revised competence* scale is .86, which provides strong evidence for internal consistency reliability. How to write about this output is found after Problems 4.4b and 4.4c.

Inter-Item Correlation Matrix

	item01 motivation	item03 competence	item05 reversed	item11 reversed
item01 motivation	1.000	.600	.761	.411
item03 competence	.600	1.000	.704	.513
item05 reversed	.761	.704	1.000	.564
item11 reversed	.411	.513	.564	1.000

Note that each of the four items in this revised competence scale correlated positively and relatively highly with the other three items. Duplicate correlations are crossed out.

See the text for details of the importance of this column

A key table.

Item-Total Statistics

	Scale Mean if Item Deleted	Scale Variance if Item Deleted	Corrected Item-Total Correlation	Squared Multiple Correlation	Cronbach's Alpha if Item Deleted
item01 motivation	9.82	5.121	.706	.589	.814
item03 competence	9.96	5.207	.718	.527	.808
item05 reversed	9.41	4.551	.831	.705	.756
item11 reversed	9.15	6.296	.558	.347	.869

If the correlation is moderately high to high (e.g., .40+), the item will make a good component of a summated rating scale. You may want to modify or delete items with low item total correlations, but in this example as are >=.40.

Interpretation of Output 4.4a continued

Third is the **Item Statistics** table of descriptive statistics, produced by checking **Item** in Figure 4.11. The fourth table is a matrix showing the **inter-item correlations** of every item in the scale with every other item. It shows whether all the items are at least moderately correlated.

The **Item Total Statistics** table, which we think is the most important table, is produced if you check **Scale if Item Deleted** under **Descriptives for** in the dialog box displayed in Figure 4.11. This table provides five pieces of information for each item in the scale. The two we find most useful are the **Corrected Item-Total Correlation** and the **Alpha if Item Deleted**. The former is the correlation of each specific item with the sum/total of the **other** items in the scale. If this correlation is moderately high or high, say, .40 or above, the item is probably at least moderately correlated with most of the other items and will make a good component of this summated rating scale. Items with lower item-total correlations do not fit into this scale as well, psychometrically. If the item-total correlation is negative or too low (less than .30), it is wise to examine the item for wording problems and conceptual fit. You may want to modify or delete such items. You can tell from the last column what the alpha would be if you deleted that item. Compare this with the alpha for the scale with all items included, which was given earlier in the **Reliability Statistics** table. Deleting a poor item will usually make the alpha go up, but it will usually make only a small difference in the alpha unless the scale has only a few items (e.g., fewer than five) because alpha is based on the number of items as well as their average intercorrelations. Note that we have used *item 05 reversed* and *item 11 reversed* so that all items would be scored with high competence as a high number. If we had instead used *item 05 low comp* or *item 11 low comp*, the item-total correlation for them probably would have been negative, indicating a problem. Some of the correlations in the **Inter-item Correlation Martix** would also be negative, which would create problems when the alpha was calculated.

Again, we ask if it is reasonable to add the item scores based on the factor analysis to form summated measures of the concepts of *motivation* and *pleasure*.

4.4b. What is the internal consistency reliability of the *revised motivation scale*?
4.4c. What is the internal consistency reliability of the *revised pleasure scale*?

Let's repeat the same steps as before to check the reliability of the following scales and then compare your output to 4.4b and 4.4c.

- For the *revised motivation scale*, use *item04 reversed, item07 motivation, item08 reversed, item12 motivation*, and *item13 motivation*.
- Remember to change the **Scale Label** to "Alpha for Revised Motivation Scale."
- Figure 4.10 should now look like Figure 4.12; the other settings are the same.

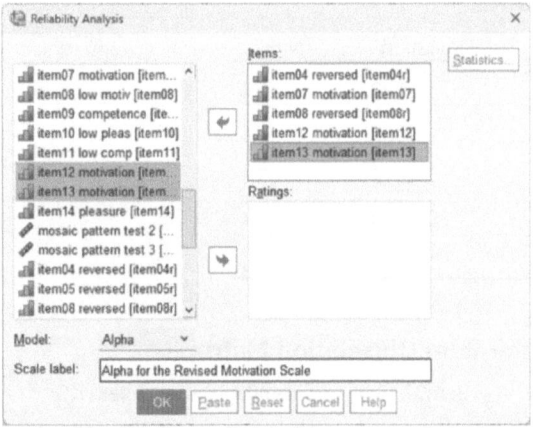

Figure 4.12 Reliability analysis.

Output 4.4b: Cronbach's Alpha for the Math Attitude Revised Motivation Scale

```
RELIABILITY
  /VARIABLES=item04r item07 item08r item12 item13
  /SCALE('Alpha for the Revised Motivation Scale') ALL
  /MODEL=ALPHA
  /STATISTICS=DESCRIPTIVE CORR
  /SUMMARY=TOTAL.
```

Reliability Scale: Alpha for the Revised Motivation Scale

Case Processing Summary

		N	%
Cases	Valid	74	98.7
	Excluded[a]	1	1.3
	Total	75	100.0

a. Listwise deletion based on all variables in the procedure.

Reliability Statistics

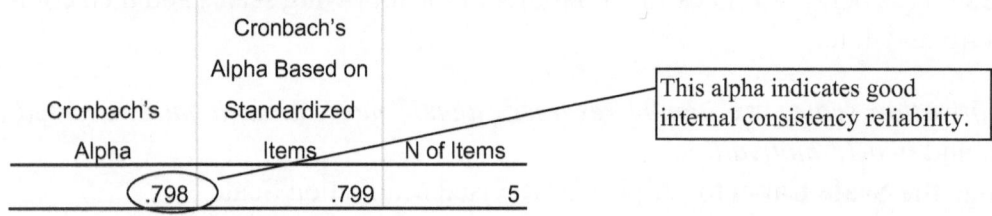

Cronbach's Alpha	Cronbach's Alpha Based on Standardized Items	N of Items
.798	.799	5

This alpha indicates good internal consistency reliability.

Item Statistics

	Mean	Std. Deviation	N
item04 reversed	2.84	.922	74
item07 motivation	2.76	1.057	74
item08 reversed	3.07	.912	74
item12 motivation	2.99	.819	74
item13 motivation	2.68	.796	74

Inter-Item Correlation Matrix

	item04 reversed	item07 motivation	item08 reversed	item12 motivation	item13 motivation
item04 reversed	1.000	.549	.584	.378	.319
item07 motivation	.549	1.000	.586	.344	.361
item08 reversed	.584	.586	1.000	.386	.314
item12 motivation	.378	.344	.386	1.000	.603
item13 motivation	.319	.361	.314	.603	1.000

Item-Total Statistics

	Scale Mean if Item Deleted	Scale Variance if Item Deleted	Corrected Item-Total Correlation	Squared Multiple Correlation	Cronbach's Alpha if Item Deleted
item04 reversed	11.49	7.404	.616	.423	.747
item07 motivation	11.57	6.824	.616	.429	.750
item08 reversed	11.26	7.372	.635	.457	.741
item12 motivation	11.34	8.145	.541	.416	.771
item13 motivation	11.65	8.395	.502	.390	.782

All of these are >.40 so good components of a revised motivation scale.

- For the *pleasure scale*, use *item02 pleasure, item06 reversed, item10 reversed,* and *item14 pleasure*.
- Change the **Scale Label** to "Alpha for the Revised Pleasure Scale."
- Figure 4.10 should now look like Figure 4.13; the other settings are the same.

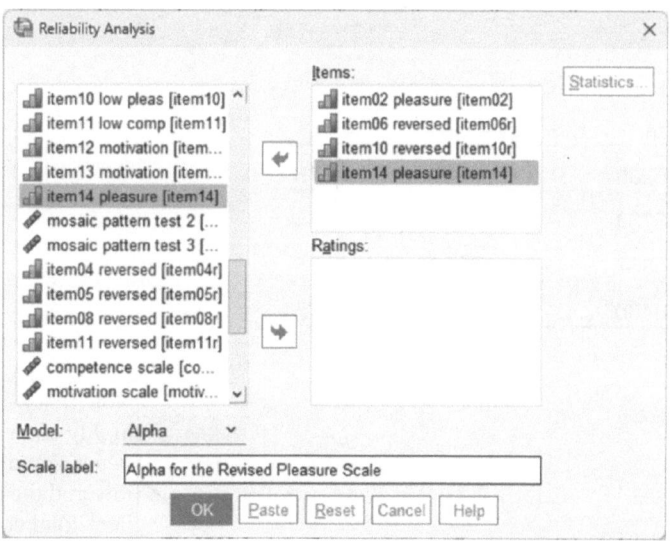

Figure 4.13 Reliability analysis.

Output 4.4c: Cronbach's Alpha for the Revised Pleasure Scale

```
RELIABILITY
  /VARIABLES=item02 item06r item10r item14
  /SCALE('Alpha for the Revised Pleasure Scale') ALL
```

```
/MODEL=ALPHA
/STATISTICS=DESCRIPTIVE CORR
/SUMMARY=TOTAL.
```

Reliability Scale: Alpha for Revised Pleasure Scale

Case Processing Summary

		N	%
Cases	Valid	75	100.0
	Excluded[a]	0	.0
	Total	75	100.0

a. Listwise deletion based on all variables in the procedure.

Reliability Statistics

Cronbach's Alpha	Cronbach's Alpha Based on Standardized Items	N of Items
.688	.704	4

This alpha is <.70 but >.60 indicates minimally acceptable internal consistency reliability.

Item Statistics

	Mean	Std. Deviation	N
item02 pleasure	3.5200	.90584	75
item06 reversed	2.5733	.97500	75
item10 reversed	3.5867	.73693	75
item14 pleasure	2.8400	.71735	75

Note that *item06* is not very highly correlated with *items02* and *10*, which lowered the alphas and the item-total correlation.

Inter-Item Correlation Matrix

	item02 pleasure	item06 reversed	item10 reversed	item14 pleasure
item02 pleasure	1.000	.285	.347	.504
item06 reversed	.285	1.000	.203	.461
item10 reversed	.347	.203	1.000	.436
item14 pleasure	.504	.461	.436	1.000

Item-Total Statistics

	Scale Mean if Item Deleted	Scale Variance if Item Deleted	Corrected Item-Total Correlation	Squared Multiple Correlation	Cronbach's Alpha if Item Deleted
item02 pleasure	9.0000	3.405	.485	.278	.615
item06 reversed	9.9467	3.457	.397	.217	.685
item10 reversed	8.9333	4.090	.407	.211	.662
item14 pleasure	9.6800	3.572	.649	.422	.528

Example of How to Write About Problems 4.4a, 4.4b, and 4.4c

Method

Based on a factor analysis of the 14 mathematics attitude items, three factors were derived. To assess whether the data from the variables in each factor form three reliable scales, Cronbach's alphas were computed. The alpha for the four item competence scale was .86, which indicates that the items would form a scale that has good internal consistency reliability. Similarly, the alpha for the motivation scale (.79) indicated good internal consistency, but the .69 alpha for the pleasure scale indicated minimally adequate reliability.

The Use of Factor Analysis and Alpha to Make Summated Scales

In Problems 4.3 and 4.4, we demonstrated how a researcher might use the results of a factor analysis to indicate how to aggregate (sum or average) the items that have high loadings for each factor and use these composite variables in further research. The implication is that each composite variable is an index of a separate underlying construct such as motivation, competence, or pleasure when studying mathematics.

It is common for a researcher to develop a smaller number of new variables from an initially larger number of items such as the 14 Likert-type ratings we designed to measure attitudes about mathematics motivation, competence, and pleasure. Figure 4.14 shows a flow chart of a method that researchers can use to help decide which items to combine or summate and how to check the internal consistency reliability of the resulting summated scales.

If the competence, motivation, and pleasure scales described in Chapter 1 had been used in other studies, we probably would have started by computing three Cronbach's alphas to check the reliability of each of the three initially planned scales: competence, motivation, and pleasure. In fact, we have done that (not shown here), and we found the Item-Total correlation for item 1 of the motivation scale was somewhat low, so we decided to use exploratory factor analyses (EFA) as shown in Problem 4.3 to determine whether the 14 items should be grouped in the way initially predicted. Based on that factor analysis we deleted item 9, and item 1 was moved from the motivation to the competence scale. Because of the changes, the alphas were recomputed for the revised grouping of items in Problem 4.4. Figure 4.14 shows the steps you could use.

Finally, the items in each group or scale should be summated or averaged to form three new composite variables. In our example, each participant's score on the four competence items would be summed to form a new composite competence variable for each person. Likewise, scores on the five motivation items would be summed, and the four low pleasure items would be summed. Now, each participant would have three new variables or measures that could be used in later inferential data analyses *instead of* their scores for the 14 original items.

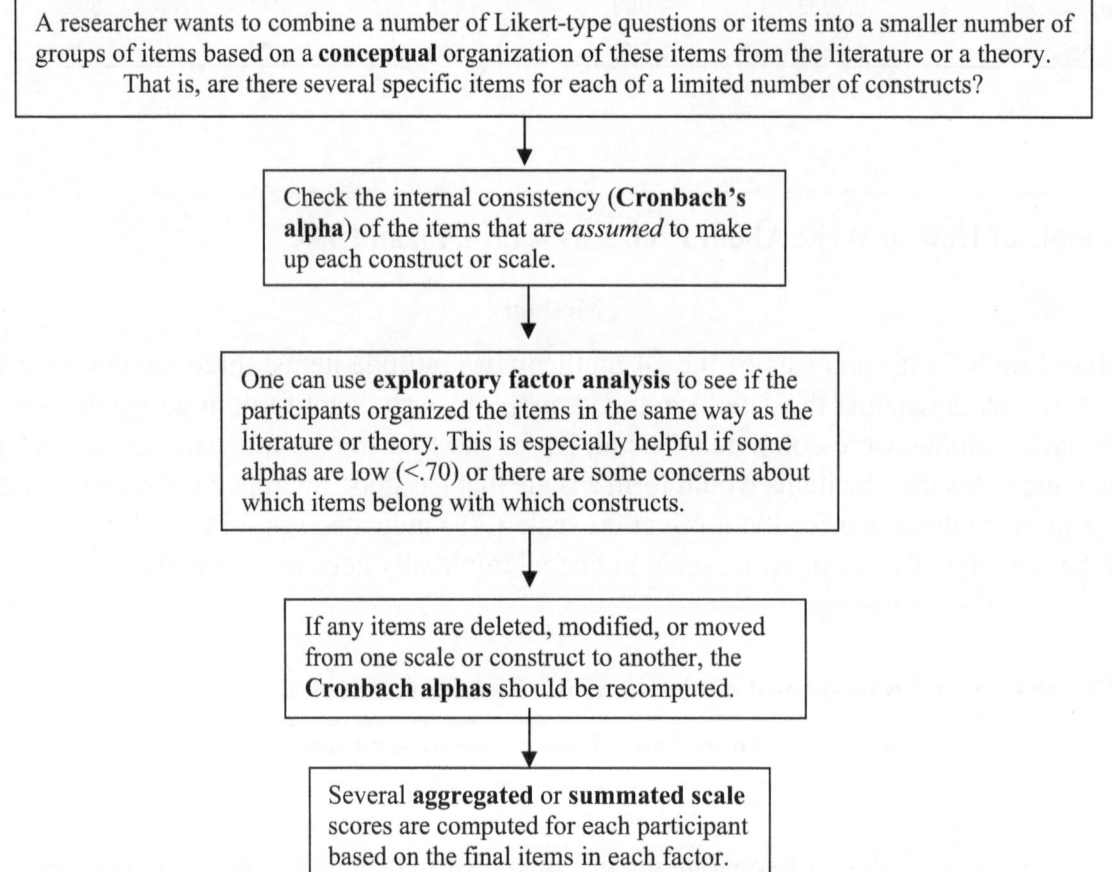

Figure 4.14 Flow chart diagram of a strategy for making multiple-item summated or composite scales when there are a number of Likert-type items that the researcher thinks can be reduced to a smaller number of conceptually meaningful scales (variables).

Interpretation Questions

4.1. Write an additional sentence or two describing the disagreements in Output 4.1 that you might include in a detailed research report.

4.2. Interpret the interrater reliability and paired *t* test results for the *visualization test* and *visualization* scores using Output 4.2. What might be another reason for the pattern of findings obtained, besides those already discussed in this chapter?

4.3. In Output 4.3 which factor explained the most variance? Why is this important?

4.4. If one were to delete one of the 14 items based on the factor analysis in Output 4.3, which one would it be? Why?

4.5. If you had to delete one question from the analysis in Output 4.4a, which one would you delete?

4.6. Interpret the Cronbach alphas in Output 4.4 a, b, and c. What is the internal consistency reliability?

Extra SPSS Problems

Using the ChapterFourData.sav file, do the following problems. Print your outputs after typing your interpretation on them. Please circle the key parts of the output that you discuss. Use the information in the appropriate "Output Interpretation" and "How to Write" sections.

4.1. Compute Cohen's kappa to assess reliability using *Rater 1* and *Rater 2*. Interpret your output as in Output 4.1 by writing an explanatory paragraph as you would in the method section of an article or paper.

4.2. Compute interrater reliability by running a paired *t* test for the *essay score 1* with the *essay score 2*. Interpret your output following the guide in the Interpretation of Output 4.2. Write up your results for this analysis as you would in the method section of an article or paper.

4.3. Run an Exploratory Factor Analysis using the extraction method of principal axis factoring as demonstrated in Problem 4.3. Run the factor analysis on the 13 items related to stress. In the extraction window (see Figure 4.7), request "fixed number of factors to extract" as 2. Study your output using the Interpretation of Output 4.3, make a table and write up your results as you would for an article or paper.

4.4. Run Cronbach's alpha for the five happiness scale items from the Chapter Six data file. Follow the procedures outlined in Problem 4.4. Write up an explanation of your resulting Cronbach's alpha finding as you would in an article or paper.

Note

1 Several other types of reliability evidence can be illustrated by this correlation. If *visualization 2* was each participant's score from retaking the test a month or so after they initially took it, then the correlation would be a measure of **test-retest reliability**. On the other hand, if *visualization 2* was a score on an alternative, parallel, or equivalent version of the *visualization test*, then this would be a measure of **equivalent forms reliability**.

5 Cross-Tabulation, Chi-Square, and Nonparametric Measures of Association

In this chapter, you will learn how to decide if there is a statistically significant relationship between two nominal variables using **chi-square**, and you will learn how to evaluate the strength of this relationship (i.e., the effect size) using **phi** (or **Cramer's V**) and **odds ratios**. You will also compute and interpret **Kendall's tau-b** for ordinal variables and **eta** for one nominal and one normal/scale variable. We will see eta again in Chapter 7 as an effect size measure for ANOVAs. The statistics demonstrated in this chapter are called nonparametric statistics because they are designed to be used with data that are not normally distributed. Many of these statistics are commonly used with data from mixed methods studies where the qualitative data have been quantitized. The statistics you will do for this chapter are found in the bottom row of Table 3.1 on selection of inferential statistics for basic or bivariate difference questions or in Table 5.1, and in the bottom row of Table 3.2 on selection of inferential statistics for bivariate association questions or in Table 5.2. We will not demonstrate the McNemar or Cochran Q Test.

- Retrieve **hsbdataNEW** from your data file. See Appendix A for how to retrieve the dataset if you need to.

Problem 5.1: Chi-Square and Phi (or Cramer's V)

The statistics discussed in this first problem are designed to analyze the association between two nominal or dichotomous variables. Remember, nominal variables are variables that have discrete *unordered* levels or categories; each subject is in only one level (e.g., students were in either the high academic track or the low academic track) of a particular nominal variable. Chi-square (χ^2) or phi/Cramer's V are good choices for statistics when analyzing the possible association between two nominal variables. They are less appropriate if either variable has three or more *ordered* levels because these statistics do not take into account the order and thus sacrifice statistical power if used with ordinal or scale variables.

Chi-square requires a relatively large sample size and/or a relatively even split of the participants among the levels because the expected counts in 80% of the cells should be greater than five. **Fisher's exact test** should be reported instead of chi-square for small samples if each of the two variables being related has only two levels (2×2 cross-tabulation), as it is more precise. Chi-square and the Fisher's exact test provide similar information about relationships among variables; however, <u>they only tell us whether the relationship is statistically significant</u> (i.e., not likely to be due to chance). <u>They do not tell the effect size</u> (i.e., the strength of the relationship). Another way to interpret chi-square is as a test of whether

DOI: 10.4324/9781003355908-5

Table 5.1

Selection of an Appropriate Inferential Statistic for Bivariate, Difference Questions or Hypotheses[a]

Scale of Measurement of **Dependent Variable**	Appropriate **Descriptive** Statistics to report	One Factor with **Two Levels** or **Categories**/Groups/Samples		One Factor **Three or More Levels** or Groups	
		Independent Samples or Groups **(Between)**	Repeated Measures or Related Samples **(Within)**	Independent Samples or Groups **(Between)**	Repeated Measures or Related Samples **(Within)**
Dependent Variable Nominal or Dichotomous	Counts, Mode, Minimum & Maximum	CHI-SQUARE	MCNEMAR	CHI-SQUARE	COCHRAN Q TEST

Table 5.2

Selection of an Appropriate Inferential Statistic for Bivariate, Associational Questions or Hypotheses

Measurement of **Both Variables**	RELATE	Two Variables or Scores for the Same or Related Subjects
Both Variables Are **Nominal** or **Dichotomous**	COUNTS OF PEOPLE	PHI or CRAMER'S *V*

there are differences between the groups formed from one variable (in this problem fast *academic track* versus regular *academic track*) in the likelihood that they are in (a) particular category(ies) of the other variable (see Table 5.1). Chi-square is telling you whether knowing a person's category on one variable systematically provides information about what category they are more likely to be in on the other variable. Chi-square is more difficult to interpret if there are more than two categories or levels in the "columns" variable so chi-square is best for 2 × 2 or 2 × 3 comparisons.

Phi (φ) and **Cramer's *V*** provide a test of statistical significance and also provide information about the *strength* of the association between two categorical variables. They can be used as measures of the effect size. If one has a 2 × 2 cross-tabulation (cross-tab), phi (φ) is the appropriate statistic. For larger cross-tabs, Cramer's *V* is used. If one variable has only two categories, then Cramer's *V* is mathematically equal to phi (φ) even if the other variable has more than two categories. The numbers in the description of the cross-tabs refer to the number of levels/categories for each of the variables. Thus, for *academic track* and *religion* in the HSB dataset, the cross-tab would be 2 × 3 because *academic track* has two levels and *religion* has three levels.

For phi and Cramer's *V*, the strength of association measures belong to the *r* family of effect sizes and are similar to the correlations you will compute in the next chapter (see Chapter 3, Table 3.5). Like correlation, a strong phi or Cramer's *V* could be close to 1.00 or −1.00, whereas one close to zero would indicate no relationship. A problem with phi and Cramer's *V* is that, under some conditions, the maximum possible value of these statistics is considerably less than 1.00. This makes them hard to interpret (see Table 5.3).

Table 5.3

Interpretation of the Strength of a Relationship (Effect Sizes) for Association Between Nominal Variables

Strength of a Relationship	Cramer's V and Phi (φ)[a]
Much larger than typical	$\geq |.70|$[b]
Large or larger than typical	$|.50|$
Medium or typical	$|.30|$
Small or smaller than typical	$|.10|$

[a] *r* family values can vary from 0.0 to + or −1.0, but except for reliability (i.e., same concept measured twice), *r* is rarely above .70. In fact, some of these statistics (e.g., phi (φ)) have a restricted range in certain cases; that is, the maximum phi is less than 1.0.

[b] Note. || indicates absolute value of the coefficient. The absolute magnitude of the coefficient, rather than its sign, is the information that is relevant to effect size.

Assumptions and Conditions for the Use of Chi-Square, Phi, Cramer's V, and Odds Ratios

1. The data for the variables must be independent. Each subject is classified into only one group on each variable and one cell of the cross-tabulation.
2. Data are treated as nominal, even if ordered.
3. For chi-square, if the expected (chance) frequencies are less than 5, the test of significance is too liberal. At least 80% of the expected frequencies should be 5 or larger. All should be at least 5 if you have a 2 × 2 chi-square; if they are not, use Fisher's exact test.
4. Odds ratios and risk ratios are problematic to interpret when the probability of an event is near zero (i.e., < .1) or near 1.

First start the Crosstabs program, which you can find under "Descriptive Statistics" even though it actually also enables you to do inferential statistics. You will first click on **Analyze → Descriptive Statistics → Crosstabs . . .** and then enter the nominal variables that you want to relate to one another. We will examine the association between students' *academic track* and whether they have mostly high or low *math grades*. To test your own research question, simply substitute your two nominal variables for *academic track* and *math grades*. In this example, we will be testing the following research questions:

5.1. Do students in fast and regular *academic tracks* differ on whether they have mainly high or low *math grades*? If so, how strong is the relationship between *academic track* and high/low *math grades*?

Let's see if students in fast and regular academic tracks differ in terms of their *math grades*. Remember, both the *academic track* and *math grades* variables are dichotomous; they each have two values.

- Click on **Analyze → Descriptive Statistics → Crosstabs . . .**
- Move *math grades* to the **Rows** box using the arrow key and put *academic track* in the **Columns** box. (Where to move these variables is a decision for the researcher. When one of the variables has more levels, it can make interpretation easier if this variable is in the Row(s) box.)

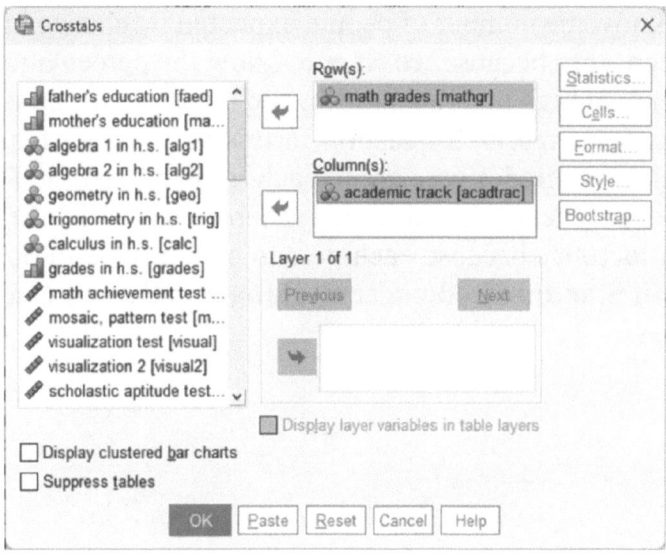

Figure 5.1 Crosstabs.

- Next, click on **Statistics** in Figure 5.1.
- Check **Chi-square** and **Phi and Cramer's V**. The window should look like Figure5.2.

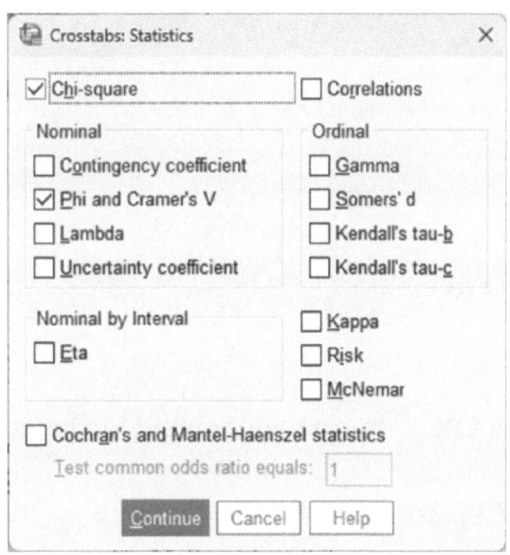

Figure 5.2 Crosstabs statistics.

- Click on **Continue**.
- Once you return to the **Crosstabs** menu (Figure 5.1), click on **Cells**.
- Now, ensure that **Observed** is checked; also in Figure 5.3, check **Expected** under **Counts**, **Column** under **Percentages**, and **Standardized** under **Residuals**. We checked **Expected**

because we want to know the number of people expected by chance in each cell. We checked **Column** under **Percentages** because we want to know the percentage of fast track and regular track students who get high and low grades, and *academic track* is the **Column** variable. We are thinking about our research question in this way, rather than as the percentage of students with high and low grades who are in each *academic track*. The total of the percentages of students for each level of the presumed independent variable (*academic track*, in this case) should add up to 100% because each person can only be in one cell of the cross-tab. Finally, we clicked on **Standardized** under **Residuals** so that we can see which cell(s) have the largest differences.

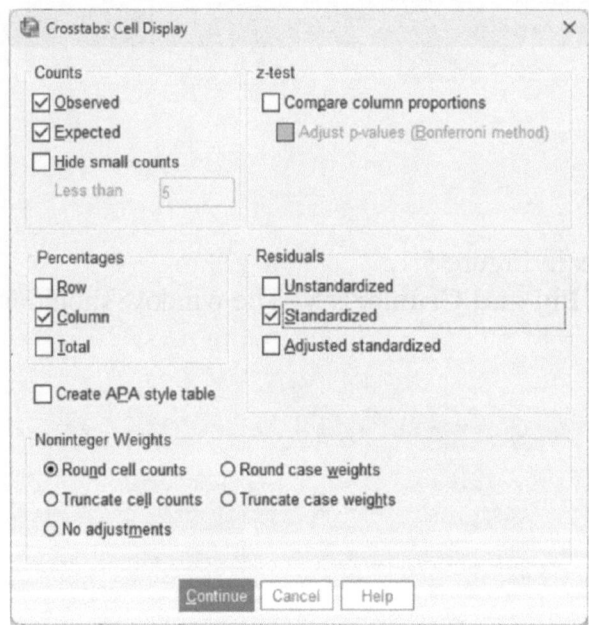

Figure 5.3 Crosstabs: Cell display.

- Click on **Continue**, then on **OK**. Compare your output to Output 5.1.

Output 5.1: Crosstabs with Chi-Square and Phi

```
CROSSTABS
  /TABLES=mathgr BY acadtrac
  /FORMAT=AVALUE TABLES
  /STATISTICS=CHISQ PHI
  /CELLS=COUNT EXPECTED COLUMN SRESID
  /COUNT ROUND CELL.
```

Crosstabs

Case Processing Summary

	Cases					
	Valid		Missing		Total	
	N	Percent	N	Percent	N	Percent
math grades * academic track	75	100.0%	0	0.0%	75	100.0%

> Compare the expected and observed counts.

math grades * academic track Crosstabulation

			academic track		Total
			fast track	regular track	
math grades	less A-B	Count	24	20	44
		Expected Count	19.9	24.1	44.0
		% within academic track	70.6%	48.8%	58.7%
		Standardized Residual	.9	-.8	
	most A-B	Count	10	21	31
		Expected Count	14.1	16.9	31.0
		% within academic track	29.4%	51.2%	41.3%
		Standardized Residual	-1.1	1.0	
Total		Count	34	41	75
		Expected Count	34.0	41.0	75.0
		% within academic track	100.0%	100.0%	100.0%

> The Standardized Residuals tell you which group has the biggest difference.

Use Pearson Chi-Square to test for significance.

	Value	df	Asymptotic Significance (2-sided)	Exact Sig. (2-sided)	Exact Sig. (1-sided)
Pearson Chi-Square	3.645[a]	1	.056		
Continuity Correction[b]	2.801	1	.094		
Likelihood Ratio	3.699	1	.054		
Fisher's Exact Test				.064	.046
Linear-by-Linear Association	3.597	1	.058		
N of Valid Cases	75				

a. 0 cells (0.0%) have expected count less than 5. The minimum expected count is 14.05.

b. Computed only for a 2x2 table

This indicates the data met the condition of having an expected count of at least 5 in each cell. If not, use the Fisher's exact test.

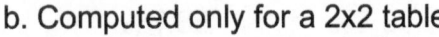

		Value	Approximate Significance
Nominal by Nominal	Phi	.220	.056
	Cramer's V	.220	.056
N of Valid Cases		75	

Best for 2x2 tables.

Interpretation of Output 5.1

The case processing summary table indicates that there are no participants with missing data. The **Crosstabulation** table includes the **Counts** and **Expected Counts,** and each cell also has a **% within academic track**. For example, there are 24 students in the fast track who had low *math grades* (less than A–B); this is 71% of the 34 fast track students. On the other hand, 20 of 41 regular track students had low math grades; that is only 49% of the regular track students. It looks like a higher percentage of fast track students had low *math grades*, but we want to make sure that this difference is systematic. The **Chi-Square Tests** table tells us whether we can be confident that this apparent difference is not due to chance.

Note, in the **Cross-tabulation** table, that the **Expected Count** of the number of fast track students who had low *math grades* is 19.9 and the observed or actual **Count** is 24. Thus, there are about 4 more fast track students who had low *math grades* than would be expected by chance, given the **Total**s shown in the table. There are also the same discrepancies between observed and expected counts in the other three cells of the table. A question answered by the chi-square test is whether these discrepancies between observed and expected counts are bigger than one might expect by chance.

The **Chi-Square Tests** table is used to determine if there is a statistically significant relationship between two dichotomous or nominal variables. It tells you whether the relationship is statistically significant but **<u>does not indicate the strength of the relationship</u>**, like a correlation, such as phi, does. In Output 5.1, we use the **Pearson Chi-Square <u>or</u>** (for small samples) the **Fisher's Exact Test** to interpret the results of the test. The two-sided (tailed) chi-square test is *not* statistically significant ($p = .056$), which indicates that we cannot be certain that fast track and regular track students are systematically different on whether they have high or low *math grades*. Note that in this case the **Fisher's Exact Test** leads to different interpretations depending on whether one uses the one-sided column ($p = .046$, statistically significant) or the two-sided column ($p = .064$, not significant). You would only use the one-sided (one-tailed) column if you had predicted *before the study* that regular track students would have higher math grades; that is, your alternative hypothesis was directional. The **Standardized Residuals** tell you which group has the biggest difference. Compare the standardized residual values in all the cells. The one with the highest absolute value has the largest difference. Also, note that footnote a states that no cells have expected counts less than 5. That is good because otherwise a condition for using chi-square would be violated. (If so, we could then use the Fisher exact test.) A good guideline is that no more than 20% of the cells should have expected frequencies less than 5. For chi-square with 1 *df* (i.e., a 2 × 2 cross-tabulation as in this case), *none* of the cells should have expected frequencies less than 5. Some statisticians say none less than 10.

The **Symmetric Measures** table provides measures of the strength of the relationships or effect size. If the association between variables is weak, the **Value** of the statistic will be relatively close to zero. If the relationship or effect size is large, the value should be +/−.50 or more. However, remember that the maximum value for phi and Cramer's *V* may be less than 1.00, the theoretical maximum for most measures of association. If both variables have two levels (i.e., 2 × 2 cross-tabs), phi is the appropriate statistic. In Output 5.1, **phi** is .22, and, like the chi-square, it is not statistically significant. Phi, in this case, is a smaller sized effect than is typical in the behavioral sciences (see Chapter 5, Table 5.4) according to Cohen (1988).

How to Write About the Results of Problem 5.1

Results

To investigate whether students in fast and regular academic tracks differ on whether they have more high or low math grades, a chi-square statistic was conducted. Assumptions were checked and were met. Table 1 shows the Pearson chi-square results and indicates that fast and regular track students are not significantly different on whether they have high or low math grades ($\chi^2 = 3.65$, $df = 1$, $N = 75$, $p = .056$). Fast track students are not more likely than expected by chance to have low or high math grades in comparison to regular track students.

Table 5.4

Chi-Square Analysis of Prevalence of High or Low Math Grades Among Fast and Regular Tracks

| Variable | n | Academic Track | | χ^2 | p |
		Fast Track	Regular Track		
Math grades				3.65	.056
Less As and Bs	44	24	20		
More As and Bs	31	10	21		
Totals	75	34	41		

Problem 5.2: Risk Ratios and Odds Ratios

When you have two dichotomous variables and thus a 2×2 contingency table or cross-tabulation, you can compute risk ratios and odds ratios instead of chi-square and phi. These ratios are commonly used to report results in medical, applied health, and prevention science fields. The assumptions for odds ratios are listed under Problem 5.1. To do odds ratios with your own data, follow the directions that follow, but substitute your variables for *math grades* and *algebra 2*. Both variables should be dichotomous (two category) variables. Again, you will use the Crosstabs program, found under **Descriptive Statistics**. For our example, we will ask:

5.2. What is the relative risk of deciding not to take algebra 2 for students having low rather than high *math grades*? And what is the odds ratio?

To compute these measures, you again use Figure 5.1, 5.2, and 5.3, but this time click on **Risk** in the **Statistics** window in Figure 5.2, to compute the risk and odds ratios.

- Click on **Analyze →Descriptive Statistics → Crosstabs . . .**
- Click on **Reset**.
- Move *algebra 2* to the **Rows** box and *math grades* to the **Columns** box (see Figure 5.1). Again, *math grades* is placed in the **Columns** box because we are thinking about this as the predictor variable (whether students have high versus low math grades).

- Click on **Statistics** to get Figure 5.2.
- Check **Risk** in the lower right part of the window. Click off **Chi-Square** and **Phi and Cramer's V** if they are checked.
- Click on **Continue** to get Figure 5.1 again.
- Click on **Cells** to get Figure 5.3 (see Problem 5.1).
- Check **Observed** and **Row**. Make sure **Expected** is not checked.
- Click on **Continue**, then **OK**.
- Compare your output and syntax to Output 5.2.

Output 5.2: Crosstabs With Risk Ratios and Odds Ratios

```
CROSSTABS
  /TABLES=alg2 BY mathgr
  /FORMAT=AVALUE TABLES
  /STATISTICS=RISK
  /CELLS=EXPECTED ROW
  /COUNT ROUND CELL.
```

Crosstabs

Case Processing Summary

| | Cases | | | | | |
| | Valid | | Missing | | Total | |
	N	Percent	N	Percent	N	Percent
algebra 2 in h.s. * math grades	75	100.0%	0	0.0%	75	100.0%

algebra 2 in h.s. * math grades Crosstabulation

| | | | math grades | | |
			less A-B	most A-B	Total
algebra 2 in h.s.	not taken	Expected Count	23.5	16.5	40.0
		% within algebra 2 in h.s.	70.0%	30.0%	100.0%
	taken	Expected Count	20.5	14.5	35.0
		% within algebra 2 in h.s.	45.7%	54.3%	100.0%
Total		Expected Count	44.0	31.0	75.0
		% within algebra 2 in h.s.	58.7%	41.3%	100.0%

> The odds ratio is the ratio of the risk of not taking algebra 2 for students with lower math grades compared to the same risk for those with higher math grades.

Risk Estimate

	Value	95% Confidence Interval	
		Lower	Upper
Odds Ratio for algebra 2 in h.s. (not taken / taken)	2.771	1.073	7.154
For cohort math grades = less A-B	1.531	1.012	2.317
For cohort math grades = most A-B	.553	.315	.970
N of Valid Cases	75		

> The risk ratio for not taking algebra 2 if you have low math grades (less A–B) is $70\% / 45.7\%$ = 1.531. See interpretation.

> The risk ratio for not taking algebra 2 if you have high math grades (most A–B) (30%/54.3%) = .553.

Interpretation of Output 5.2

The first two tables are similar to those in Output 5.1, except that the variables are different and there are no **Expected Counts** in the **Crosstabulation** table. Note that all 75 participants had both variables. Forty-four students had low *math grades* (few As and Bs). Of those 44, 28 did not take *algebra 2* and 16 did take it. The **Risk Estimate** table shows the odds ratio, the two risk ratios, and confidence intervals for each. The first risk ratio is 1.53, which is computed by dividing 70% (percentage of students with low math grades who did not take algebra 2) by 45.7% (percentage of students with low math grades who did take algebra 2). This risk ratio means that students who had low *math grades* were about 1½ times as likely to **not** take *algebra 2* as they were to take *algebra 2*. Conversely, of the 31 students with high math grades (mostly As and Bs), 12 didn't take *algebra 2* and 19 did. The second risk ratio is .553, which is 30% (percentage with high math grades who didn't take algebra 2) divided by 54.3% (percentage with high math grades who did take it). This risk ratio is interpreted as students with high *math grades* are a little over half as likely to **not** take *algebra 2* as to take it. Notice that the **95% Confidence Interval** for each ratio does not include 1.0. That is, the **Lower and Upper** bounds are either greater than 1.0 (i.e., 1.012 and 2.317) or less than 1.0 (.315 and .970). This indicates that the risk ratios are statistically significant. If the risk ratio was exactly 1.0, it would indicate that, for example, those who have low *math grades* are equally likely to have taken *algebra 2* as they are not to have taken *algebra 2*.

The **Odds Ratio** (OR) of 2.77 is a ratio of ratios. In this case, 1.531/.553 = 2.77. This OR means that the odds of failing to take *algebra 2* are 2.77 times higher for those who had low *math grades* as for those who had high *math grades*.

Odds ratios and risk ratios are common examples of a third group or family of effect size measures, called **risk potency** measures. Remember that we discussed in Chapter 6 the *r* (relationship) family of effect sizes, including phi, Cramer's *V*, and eta calculated in this chapter, and the *d* (differences) family, which is used to indicate effect size in Chapter 7. Although odds ratios and risk ratios are common effect size measures when both variables are dichotomous (also called binary), especially in the health-related literature, there are no agreed-upon standards for what represents a large ratio because the ratio may approach infinity if the outcome is very rare or very common, even when the association is near random.

How to Write About Output 5.2

Results

Because whether or not students took algebra 2 and whether their math grades were high or low were both binary variables, and neither alternative was rare, an odds ratio was computed. The OR was 2.77, indicating that the odds of students failing to take algebra 2 were 2.77 times higher if they had low math grades compared to if they had high math grades. The 95% confidence interval was 1.07 to 7.15.

Problem 5.3: Other Nonparametric Associational Statistics

In addition to phi and Cramer's *V*, there are several other nonparametric measures of association that we could have chosen in Figure 5.2. They attempt, in different ways, to measure the strength of the association between two variables. If both variables are nominal and you have a 2 × 2 cross-tabulation, like the one in Output 5.1, phi is the appropriate statistic to use from the symmetric measures table. For larger cross-tabulations (like a 3 × 3) with nominal data, Cramer's *V* is the appropriate statistic. Note that with a 2 × 3 or a 3 × 2 cross-tabulation, phi and Cramer's *V* are the same. If the variables are ordered (i.e., ordinal), however, you have several other choices. We will use Kendall's tau-b in this problem. The primary assumption of Kendall's tau-b is that data are at least ordinal.

5.3. What is the relationship or association between *father's education revised* and *mother's education revised*?

- **Analyze → Descriptive Statistics → Crosstabs . . .**
- Click on **Reset** to clear the previous entries.
- Put *mother's education revised* in the **Rows** box and *father's education revised* in the **Columns** box.
- Click on **Cells** and ask the **Observed** and **Expected** cell counts and **Total** percentages be printed in the table. Click on **Continue** and then **Statistics**.
- Request the following **Statistics: Kendall's tau-b** coefficient under **Ordinal**, and **Phi and Cramer's *V*** under **Nominal** (for comparison purposes). Do not check **Chi-square**.
- Click on **Continue** then **OK**. Compare your syntax and output to Output 5.3.

Output 5.3: Crosstabs and Nonparametric Associational Statistics

```
CROSSTABS
  /TABLES=maedRevis BY faedRevis
  /FORMAT=AVALUE TABLES
  /STATISTICS=PHI BTAU
  /CELLS=COUNT EXPECTED TOTAL
  /COUNT ROUND CELL.
```

Crosstabs

Case Processing Summary

	Cases					
	Valid		Missing		Total	
	N	Percent	N	Percent	N	Percent
mother's educ revised * father's educ revised	73	97.3%	2	2.7%	75	100.0%

mother's educ revised * father's educ revised Crosstabulation

			father's educ revised			
			HS grad or less	Some College	BS or More	Total
mother's educ revised	HS grad or less	Count	33	9	4	46
		Expected Count	23.9	10.1	12.0	46.0
		% of Total	45.2%	12.3%	5.5%	63.0%
	Some College	Count	5	7	7	19
		Expected Count	9.9	4.2	4.9	19.0
		% of Total	6.8%	9.6%	9.6%	26.0%
	BS or More	Count	0	0	8	8
		Expected Count	4.2	1.8	2.1	8.0
		% of Total	0.0%	0.0%	11.0%	11.0%
Total		Count	38	16	19	73
		Expected Count	38.0	16.0	19.0	73.0
		% of Total	52.1%	21.9%	26.0%	100.0%

> Phi is not appropriate for a 3 × 3 table.

> Cramer's *V* measures the strength of a relationship of two nominal variables when one or both have three or more levels/values.

Symmetric Measures

		Value	Asymptotic Standard Error[a]	Approximate T[b]	Approximate Significance
Nominal by Nominal	Phi	.710			<.001
	Cramer's V	.502			<.001
Ordinal by Ordinal	Kendall's tau-b	.572	.084	5.835	<.001
N of Valid Cases		73			

a. Not assuming the null hypothesis.

b. Using the asymptotic standard error assuming the null hypothesis.

> Kendall's tau-b measures the strength of the association if both variables are ordinal.

Interpretation of Output 5.3

There are several nonparametric measures of association that we could have chosen from Figure 5.2. All of them except chi-square attempt, in different ways, measure the *strength* of the association between two variables roughly on the –1 to +1 scale used by the Pearson correlation (see Chapter 6). However, several of them, including phi and Cramer's *V*, have maximum values considerably less than 1 under some conditions.

For tables **with nominal data and 3 levels/groups or more for both variables** (like a 3 × 3 cross-tabulation of *religion* and *ethnicity*), **Cramer's *V*** would be the appropriate statistic. However, in Problem 5.3, we requested **Kendall's tau-b** because both *mother's education* and *father's education* are ordered variables (ordinal data). Cramer's *V* (and phi) treat the cross-tabulated variables as if they were nominal even if they are ordered, so they are actually not good choices for this problem. We checked that box so you could compare Cramer's *V* to Kendall's tau-b. Phi is not appropriate for a relation between two variables that have three categories each.

If the association between variables is weak, the value of the statistic will be close to zero, and the **Approximate Significance** level will be greater than .05, the usual cutoff to say that an association is statistically significant. However, if the association is statistically significant, the *p* will be small ($< .05$). In this case, *p* is $< .001$ for Kendall's tau-b, which is clearly significant, and the effect size (tau-b = .572) is large; the interpretation of tau-b is similar to that of *r* (see Chapter 5, Table 5.5).

Example of How to Write About Problem 5.3

Results

To investigate the relationship between father's education and mother's education, Kendall's tau-b was conducted. The analysis indicated a statistically significant positive association between father's education and mother's education, tau (71) = .572, $p < .001$. This means that more highly educated fathers were married to more highly educated mothers, and less educated fathers were married to less educated mothers. This tau is considered to be a large effect size (Cohen, 1988).

Problem 5.4: Eta

There is an important associational statistic, eta, that is used when one variable is nominal and the other is approximately normal or scale. We will use this statistic to describe the association between *academic track* and *math courses taken* (an approximately normal variable with six levels). **Eta squared** will be an important statistic in later chapters when we interpret the *effect size* of various ANOVAs.

5.4. What is the association between *academic track* and number of *math courses taken*? How strong is it?

Follow these steps:

- Click on **Analyze → Descriptive Statistics → Crosstabs ...**
- Click on **Reset** to clear the previous entries.

- Put *math courses taken* in the **Rows** box using the arrow key and put *academic track* in the **Columns** box (similar to Figure 5.1), again because you are thinking of academic track as the independent/grouping variable.
- Next, click on **Statistics** and select **Eta**.
- Click on **Continue**.
- Now, click on **Cells** and select **Expected** and **Observed**.
- Click on **Continue**.
- Click on **OK**. Compare your syntax and output to Output 5.4.

Output 5.4: Eta for Academic Track and Math Courses Taken

```
CROSSTABS
  /TABLES=mathcrs BY acadtrac
  /FORMAT=AVALUE TABLES
  /STATISTICS=ETA
  /CELLS=COUNT EXPECTED
  /COUNT ROUND CELL.
```

Crosstabs

Case Processing Summary

	Cases					
	Valid		Missing		Total	
	N	Percent	N	Percent	N	Percent
math courses taken * academic track	75	100.0%	0	0.0%	75	100.0%

math courses taken * academic track Crosstabulation

			academic track		Total
			fast track	regular track	
math courses taken	0	Count	4	12	16
		Expected Count	7.3	8.7	16.0
	1	Count	3	13	16
		Expected Count	7.3	8.7	16.0
	2	Count	9	6	15
		Expected Count	6.8	8.2	15.0
	3	Count	6	2	8
		Expected Count	3.6	4.4	8.0
	4	Count	7	5	12
		Expected Count	5.4	6.6	12.0
	5	Count	5	3	8
		Expected Count	3.6	4.4	8.0
Total		Count	34	41	75
		Expected Count	34.0	41.0	75.0

Directional Measures

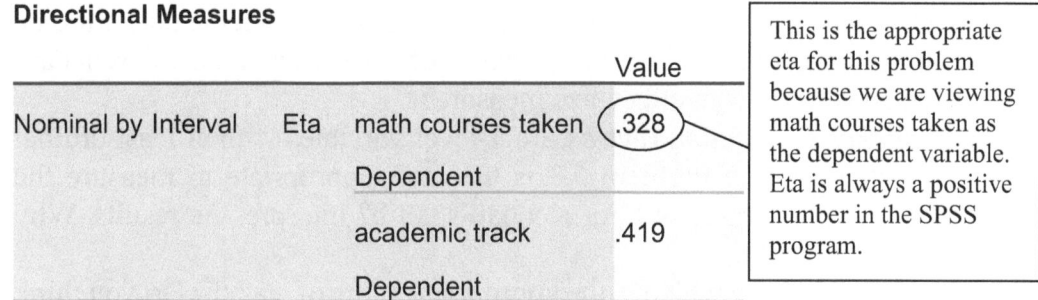

			Value
Nominal by Interval	Eta	math courses taken Dependent	.328
		academic track Dependent	.419

> This is the appropriate eta for this problem because we are viewing math courses taken as the dependent variable. Eta is always a positive number in the SPSS program.

Interpretation of Output 5.4

The second table shows the actual **Counts** and the **Expected Counts** of the number of persons in each cell. If there are positive discrepancies between the actual and expected counts in the upper left (fast track) columns and negative discrepancies in the lower left columns or vice versa, that would indicate that there is an association between the two variables. Because of the way the SPSS program computes eta, it ranges from zero to about +1.0. High values of eta indicate a strong association. In this case, the appropriate eta is .328 because *math courses taken* is the dependent variable. It is a medium to large effect size (see Chapter 3, Table 3.5). With 75 subjects, an eta of .33 probably would be statistically significant, but this program does not test for significance. Eta squared would be .11, indicating that the two variables share 11% common variance. We will see eta squared when interpreting the size of the "effect" in analysis of variance.

Example of How to Write About Problem 5.4

Results

Eta was used to investigate the strength of the association between academic track and number of math courses taken (eta = .33). This is a medium to large effect size (Cohen, 1988). Those in the fast track were more likely to take several or all the math courses than those in the regular track.

Interpretation Questions

5.1. In Output 5.1: (a) What do the terms "count" and "expected count" mean? (b) What does the difference between them tell you?

5.2. In Output 5.1: (a) Is the (Pearson) chi-square statistically significant? Explain what it means. (b) Are the expected values in at least 80% of the cells $\geq$ 5? How do you know? Why is this important?

5.3. In Output 5.2: (a) How is the risk ratio calculated? What does it tell you? (b) How is the odds ratio calculated and what does it tell you? (c) How could information about the odds ratio be useful to people wanting to know the practical importance of research results? (d) What are some limitations of the odds ratio as an effect size measure?

5.4. Because *father's* and *mother's education revised* are 3-level variables with at least ordinal data, which of the statistics used in Problem 5.3 is the most appropriate to measure the strength of the relationship: phi, Cramer's *V*, or Kendall's tau-b? Interpret the results. Why are tau-b and Cramer's *V* different?

5.5. In Output 5.4: (a) How do you know which is the appropriate value of eta? (b) Do you think it is high or low? Why? (c) How would you describe the results?

Extra SPSS Problems

Using the CollegeStudentData.sav file, do the following problems. Print your outputs after typing your interpretations on them. Please circle the key parts of the output that you discuss.

5.1. Run crosstabs and interpret the results of chi-square and phi (or Cramer's *V*), as discussed in Chapter 5 and in the interpretation of Output 5.1, for: (a) academic track and marital status and (b) age group and marital status.

5.2. Select two other appropriate variables; run and interpret the output as we did in Output 5.1.

5.3. Is there an association between having children or not and watching TV sitcoms?

5.4. Is there a difference between students who have children and those who do not in regard to their age group?

5.5. Compute an appropriate statistic and effect size measure for the relationship between academic track and evaluation of social life.

6 Correlation and Regression

In this chapter, you will learn how to compute several associational statistics, after you learn how to make **scatterplots** and how to interpret them. An assumption of the **Pearson product moment correlation (*r*)** is that the variables are related in a linear (straight line) way so we will examine scatterplots to see if that assumption is reasonable. Second, the **Pearson *r*** and the **Spearman rho (*r_s*)** will be computed. The Pearson *r* is used when you have two variables that are normal/scale, and the Spearman rho is used when one or both of the variables are ordinal or if either or both variable is not normally distributed. Third, you will compute a **correlation matrix** indicating the associations among all the pairs of three or more variables. Fourth, you will compute simple or **bivariate (2 variable) regression**, which is used when one wants to predict scores on a normal/scale dependent (outcome) variable from one normal or scale independent (predictor) variable. Last, we will provide an introduction to a complex associational statistic, **multiple regression**, which is used to predict a scale/normal dependent variable from two or more independent variables.

As stated in Chapter 3, correlations can vary from –1.00 (a perfect negative correlation or association) through .00 (no correlation) to +1.0 (a perfect positive correlation). Note that +1 and –1 are equally high or strong, but they lead to different interpretations. A high **positive correlation** between aptitude and grades would mean that students with higher aptitude tended to have higher grades, those with lower aptitude had lower grades, and those in between had grades that were neither especially high nor especially low. A high **negative correlation** would mean that students with higher aptitude tended to have lower grades; as student excel more and more in aptitude, they get lower and lower grades. With a **zero correlation** there are no consistent associations. A student with high aptitude might have low, medium, or high grades.

Assumptions and Conditions for the Pearson Correlation (*r*) and Bivariate Regression

1. The two variables have a linear relationship. We will show how to check this assumption with a scatterplot in Problem 6.1. (Pearson *r* will not detect a curvilinear relationship unless you transform the variables, which is beyond the scope of this book.)
2. Scores on one variable are normally distributed for each value of the other variable and vice versa. If degrees of freedom are greater than 25, failure to meet this assumption has little consequence. Statistics designed for normally distributed data are called **parametric statistics**. Remember that we checked variables for skewness and normality in Chapter 2.
3. Outliers (i.e., extreme scores) can have a big effect on the correlation.

DOI: 10.4324/9781003355908-6

Spearman Rho is computed by ranking the scores for each variable and then computing a Pearson product moment correlation on the ranks. The SPSS program will rank the variables for you automatically when you request a Spearman correlation, so there is no need to rank them yourself.

Assumptions and Conditions for Spearman Rho (r_s)

1. Data on both variables are at least ordinal. Statistics designed for ordinal data and which do not assume normal distribution of data are called **nonparametric statistics**.
2. Scores on one variable are monotonically related to the other variable. This means that as the values of one variable increase, the other should also increase but not necessarily in a linear (straight line) fashion. The curve can flatten but cannot go both up and down as in a **U** or **J**.

Table 6.1 provides a flowchart for selecting association statistics. It is separated into two parts: The first part shows how to select a statistic when you want to simply relate two variables to one another, without viewing either as a predictor/independent or dependent variable. The second part helps you select a statistic that will enable you to predict one dependent variable from one or more predictors/independent variables.

Table 6.1

Selection of an Appropriate Inferential Statistic for Associational Questions or Hypotheses

Bidirectional Association Between Two Variables		
Measurement of **Both Variables**	RELATE	Two Variables or Scores for the Same or Related Subjects
Variables Are Both **Normal/Scale** and Assumptions Not Markedly Violated	SCORES	PEARSON (r)
Both Variables at Least **Ordinal** Data or Assumptions Markedly Violated	RANKS	SPEARMAN (Rho)

Prediction of One Normal/Scale Dependent Variable from One or More Independent or Predictor Variables			
One Normal/Scale Independent Variable	Several Independent or Predictor Variables		
	Normal or Scale	Some Normal Some Dichotomous (Two category)	All Dichotomous
BIVARIATE REGRESSION	MULTIPLE REGRESSION	MULTIPLE REGRESSION	MULTIPLE REGRESSION

Table 6.2

Interpretation of the Strength of a Relationship (Effect Sizes) for Association Between Two Normal/Scale or Ordinal Variables

Strength of a Relationship	r, rho [a]
Much larger than typical	$\geq |.70|$ [b]
Large or larger than typical	$|.50|$
Medium or typical	$|.30|$
Small or smaller than typical	$|.10|$

[a] *r* family values can vary from 0.0 to + or −1.0, but except for reliability (i.e., same concept measured twice), *r* is rarely above .70.

[b] Note. | | indicates absolute value of the coefficient. The absolute magnitude of the coefficient, rather than its sign, is the information that is relevant to effect size.

• Retrieve **hsbdataNEW**. See Appendix A for how to retrieve the dataset if you need to.

Problem 6.1: Scatterplots to Check the Assumption of Linearity

A **scatterplot** is a plot or graph of two variables that shows whether and how the score for an individual on one variable is related to their score on the other variable. If the correlation is *high positive*, the plotted points will be close to a straight line (the **linear regression line**) from the lower left corner of the plot to the upper right. The linear regression line will slope downward from the upper left to the lower right if the correlation is *high negative*. For correlations *near zero*, the regression line will be flat with many points far from the line, and the points form a pattern more like a circle or random blob than a line or oval.

On the scatterplot, each dot or circle on the plot represents a particular individual's score on the two variables, with one variable being represented on the X axis and the other on the Y axis. The plot also allows you to see if there are bivariate outliers (circles/dots that are far from the regression line, indicating that the way that person's score on one variable relates to his/her score on the other is different from the way the two variables are related for most of the other participants), and it may show that a better fitting line would be a curve rather than a straight line. In this case the assumption of a linear relationship is violated and a Pearson correlation would not be the best choice. The Spearman or Kendall's tau-b correlations would be better.

6.1. What are the scatterplots and linear regression line for (a) *math achievement* and *grades in h.s.* and for (b) *math achievement* and *mosaic pattern score*?

To develop a scatterplot of *math achievement* and *grades*, follow these commands:

• **Graphs → Scatter/Dot**. This will give you Figure 6.1.
• Click on **Simple Scatter**.

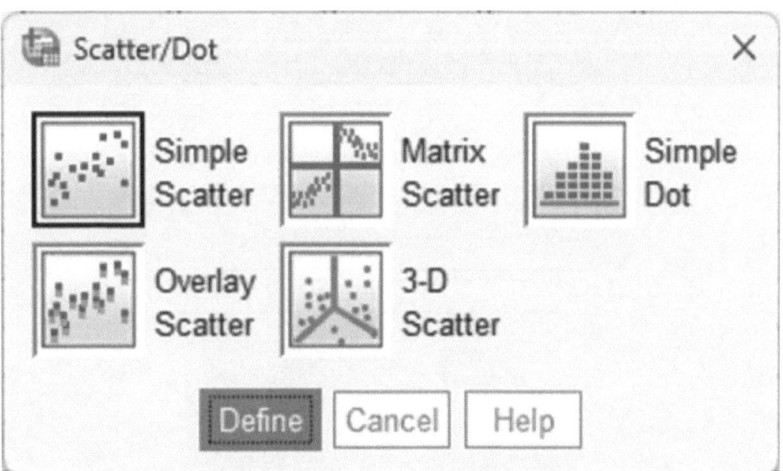

Figure 6.1 Scatterplot.

• Click on **Define**, which will bring you to Figure 6.2.
• Now, move *math achievement* to the **Y Axis** and *grades in h.s.* to the **X Axis**. Note: the presumed outcome or dependent variable goes on the Y axis. However, for the correlation itself, there is no real independent or dependent variable; both variables are treated the same.

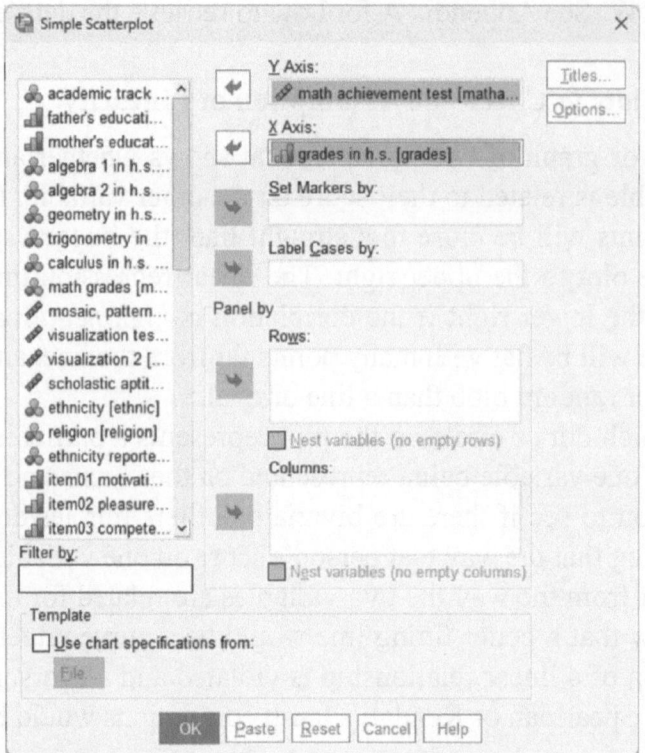

Figure 6.2 Simple scatterplot.

- Next, click on **Titles** (in Figure 6.2) so that you will know what the scatterplot represents. Type **Correlation of math achievement with high school grades** (see Figure 6.3).

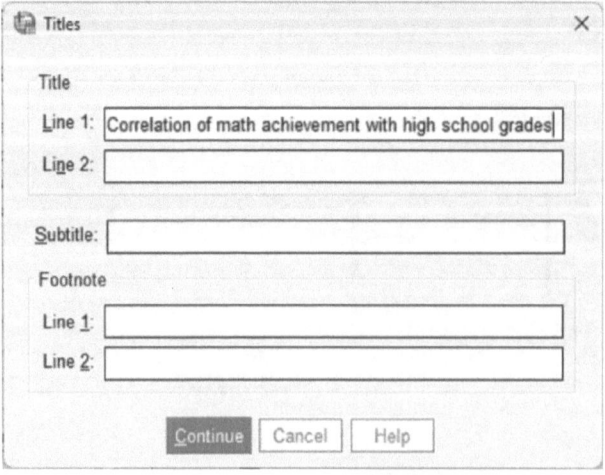

Figure 6.3 Titles.

- Click on **Continue**, then on **OK**. You will get Output 6.1a, the scatterplot. Before we interpret this, we will add the regression line in order to get a better sense of the linear relationship and how much scatter or deviation around that line there is.

Output 6.1a: Scatterplot Without Regression Line

```
GRAPH
  /SCATTERPLOT(BIVAR)=grades WITH mathach
  /MISSING=LISTWISE
  /TITLE= 'Correlation of math achievement with high school grades'
```

Graph

Scatterplots result from plotting points on a graph. Each circle represents <u>one</u> participant's score on the two variables. The pattern indicates the strength and direction of the association between the two variables. Note that the points line up along the only possible numbers for high school grades. This "lining up" can be ignored; just look at the overall direction of the points. Later, in 7.1b, look at how close they are to the regression line

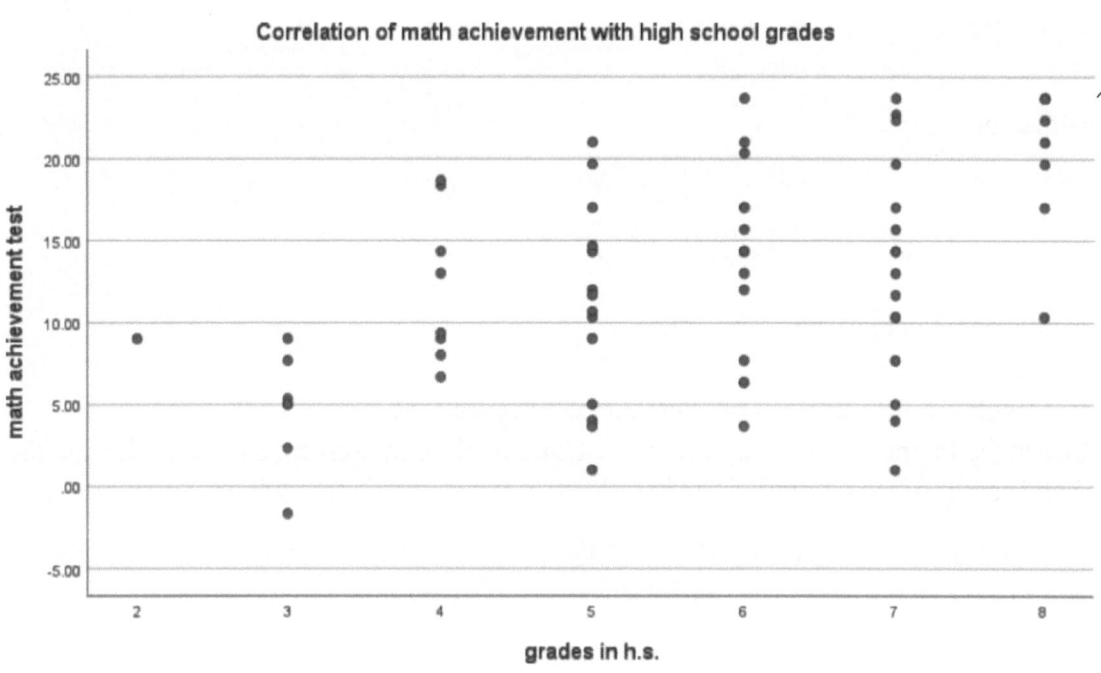

- <u>Double click on the scatterplot in Output 6.1a</u>. The Chart Editor (Figure 6.4) will appear as a new window.
- Click on the button circled in Figure 6.4 to create a **Fit Line**. The **Properties** window (see Figure 6.5) will appear as well as a fit line in the Chart Editor.

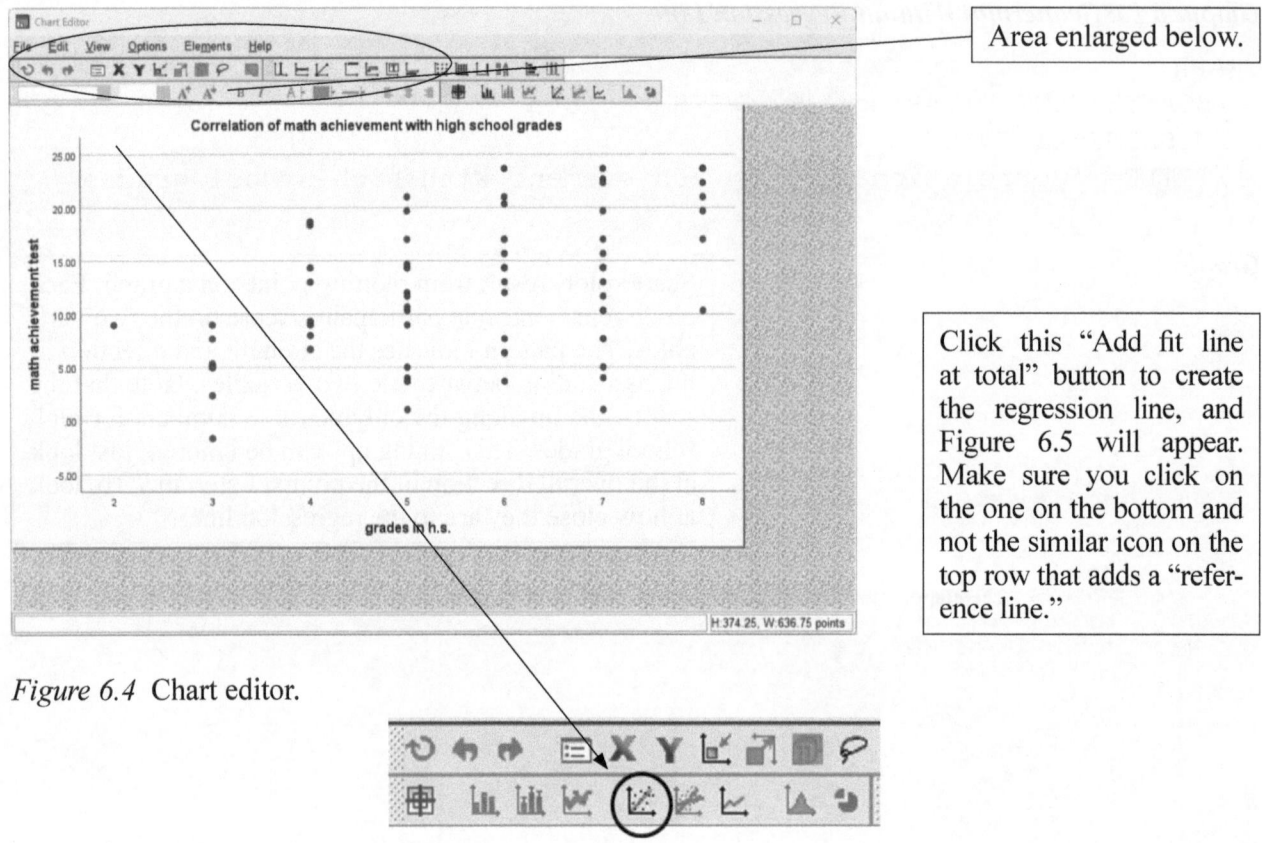

Area enlarged below.

Click this "Add fit line at total" button to create the regression line, and Figure 6.5 will appear. Make sure you click on the one on the bottom and not the similar icon on the top row that adds a "reference line."

Figure 6.4 Chart editor.

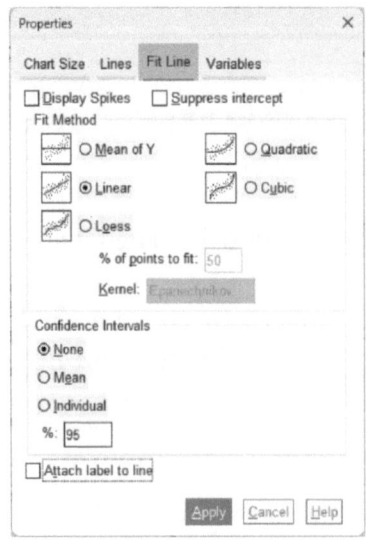

Illustration 6.1 Enlarged Figure of the Chart Editor

- Be sure that **Attach label to line** is not checked (see Figure 6.5).
- Click on **Close** in the **Properties** window and click **Close** to close the screen, then click the x in the upper right hand corner of the **Chart Editor** in order to return to the Output window (Output 6.1b).

Figure 6.5 Properties.

Output 6.1b: Scatterplot With Regression Lines

```
GRAPH
  /SCATTERPLOT(BIVAR)=grades WITH mathach
  /MISSING=LISTWISE
  /TITLE= 'Correlation of math achievement with' 'high school
  grades'.
```

Graph

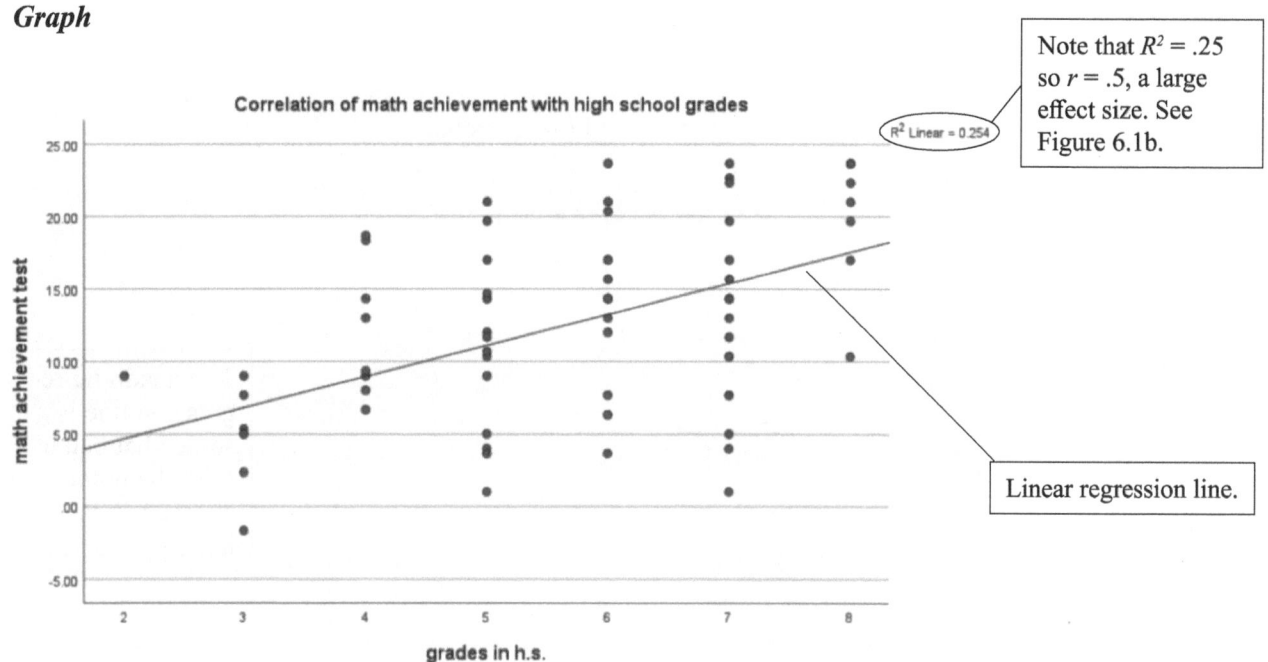

Note that $R^2 = .25$ so $r = .5$, a large effect size. See Figure 6.1b.

Linear regression line.

• Now <u>add a new scatterplot</u> to Output 6.1b by doing the same steps that you used for Problem 6.1a **Graphs → Scatter/dot** <u>for a new pair of variables</u>: *math achievement* (**Y Axis**) with *mosaic* (**X Axis**).
• Don't forget to click on **Titles** and change it before you run the scatterplot so that the title reads: **Correlation of math achievement with mosaic pattern score**.
• Click **Continue** and **OK.**
• <u>Now you need to add a Quadratic and a Linear regression line to the chart</u>. Double click on the scatterplot in Output 6.1b. The Chart Editor (Figure 6.4) will appear.
• Click on the button circled in Figure 6.4 to create a **Fit Line**. The **Properties** window (see Figure 6.5) will appear as well as a fit line in the Chart Editor.
• Click **Quadratic** and ensure **Attach label to line** is not checked.
• Click **Apply** and **Close.**
• Click the button circled in Figure 6.4 again.
• Click **Linear** and ensure **Attach label to line** is not checked.
• Click **Apply** and **Close** then close the Chart Editor by clicking the x in the upper right corner.

Do your syntax and scatterplots look like the ones in Output 6.1c?

Output 6.1c: Output With Linear and Quadratic Regression Lines

```
GRAPH
  /SCATTERPLOT(BIVAR)=mosaic WITH mathach
  /MISSING=LISTWISE
  /TITLE= 'Correlation of math achievement with mosaic pattern
  score'.
```

Graph

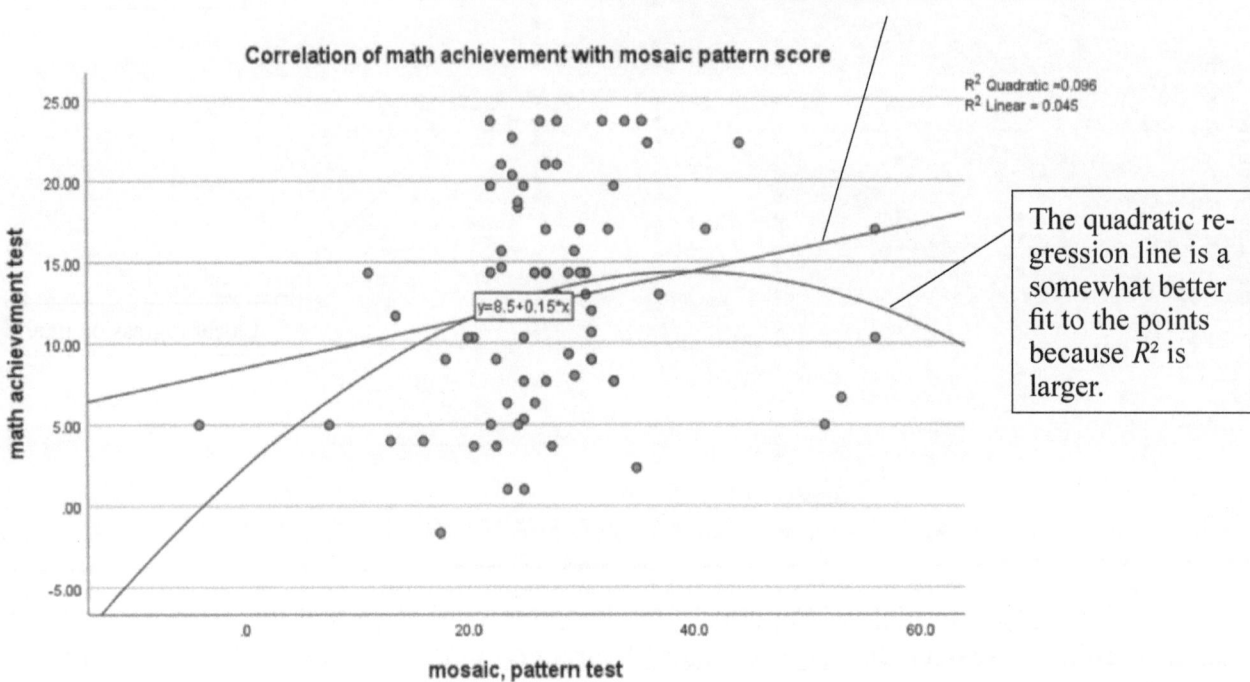

The linear regression line. Note that, in contrast to the previous plot, many points are not near the line, and the points form a cloud, rather than clustering along the line. This plot shows a poor fit and low correlation.

The quadratic regression line is a somewhat better fit to the points because R^2 is larger.

Interpretation of Outputs 6.1b and 6.1c

Both scatterplots shown in Outputs 6.1b and 6.1c show the straight or linear regression line that has "best fit" (i.e., it minimizes the squared differences between the points and the line). Note that for the first scatterplot (grades in h.s.), the points fit the line pretty well; $r^2 = .25$ and thus r is .50. The second scatterplot shows that mosaic and math achievement are only weakly correlated; the points do not fit the line very well, $r^2 = .05$, and r is .21. In the second scatterplot we asked the program to fit a quadratic (one bend) curve as well as a linear line. It seems to fit the points better; $r^2 = .10$. This suggests that the linear assumption is violated and a Pearson correlation may not be the most appropriate statistic.

Problem 6.2: Bivariate Pearson and Spearman Correlations

The **Pearson product moment correlation** is a bivariate parametric statistic used when both variables are approximately normally distributed (i.e., scale data). When you have ordinal data or when assumptions are markedly violated, one should use a nonparametric equivalent of the Pearson correlation coefficient. One such the **Spearman rho** (another is Kendall's tau-b). Here you will compute both parametric and nonparametric correlations and then compare them. The variables of interest for Problem 6.2 are *mother's education* and *math achievement*. We found in Chapter 2 that *mother's education* was somewhat skewed, but that *math achievement* was normally distributed. Ordinarily, we would not calculate both Pearson and Spearman correlations on the same two variables. We would calculate Pearson if both variables are reasonably normally distributed and Spearman if one or both variables is ordinal and/or not normally distributed. We do both here to show you how the two types of correlation compare to one another. These two types of correlations involve the same formula, but the Pearson applies it to raw, untransformed scores, and the Spearman applies it to ranks of the scores.

6.2. What is the association between *mother's education* and *math achievement*?

To compute Pearson and Spearman correlations follow these commands:

- **Analyze → Correlate → Bivariate . . .**
- Move *math achievement* and *mother's education* to the **Variables** box.
- Next, under **Correlation Coefficients**, ensure that the **Spearman** and **Pearson** boxes are checked.
- Make sure that the **Two-tailed** (under **Test of Significance**) and **Flag significant correlations** are checked (see Figure 6.6). Unless one has a clear directional hypothesis, two-tailed tests are used. Flagging the significant correlations (with an asterisk) is optional, but it helps you quickly identify the statistically significant correlations.

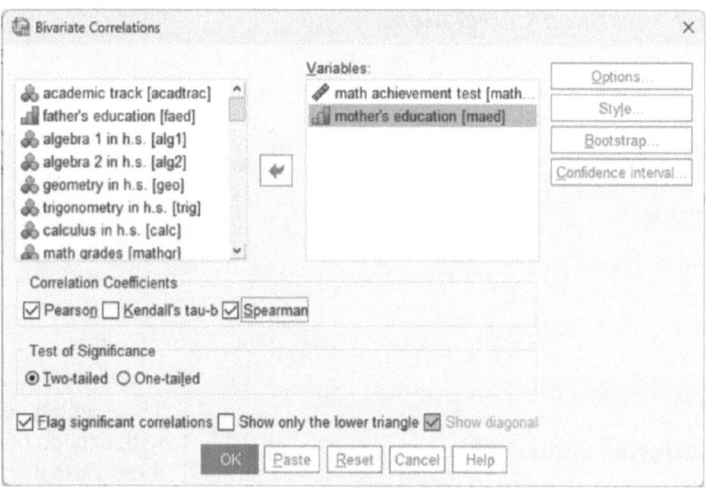

Figure 6.6 Bivariate correlations.

- Now click on **Options** to get Figure 6.7.

- Click on **Means and standard deviations** so that you will have descriptive statistics on the variables for the participants who are included in the correlations, and click on **Exclude cases listwise**. "Listwise exclusion" means that if a person is missing data for any of the variables in the correlation matrix, that person will be excluded from all of the correlations. That way, you can compare the correlations and know that they include exactly the same people. When you request only one correlation, listwise and pairwise exclusion (of participants with missing data on one or both variables) are the same, but as described later, which one you select may make a difference in a correlation matrix of more than one pair of variables.

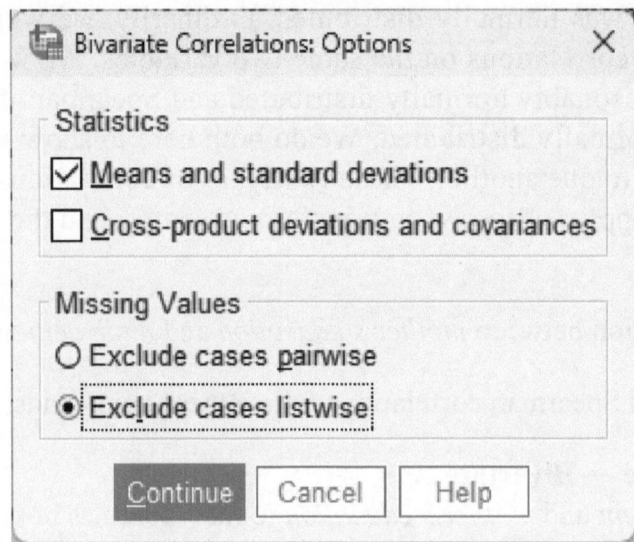

Figure 6.7 Bivariate correlations: Options.

- Click on **Continue** then on **OK**. Compare Output 6.2 to your output and syntax.

Output 6.2: Pearson and Spearman Correlations

```
CORRELATIONS
  /VARIABLES=mathach maed
  /PRINT=TWOTAIL NOSIG
  /STATISTICS DESCRIPTIVES
  /MISSING=LISTWISE.
```

Correlations

Descriptive Statistics

	Mean	Std. Deviation	N
math achievement test	12.5645	6.67031	75
mother's education	4.11	2.240	75

> There are 75 persons with data on both of these variables.

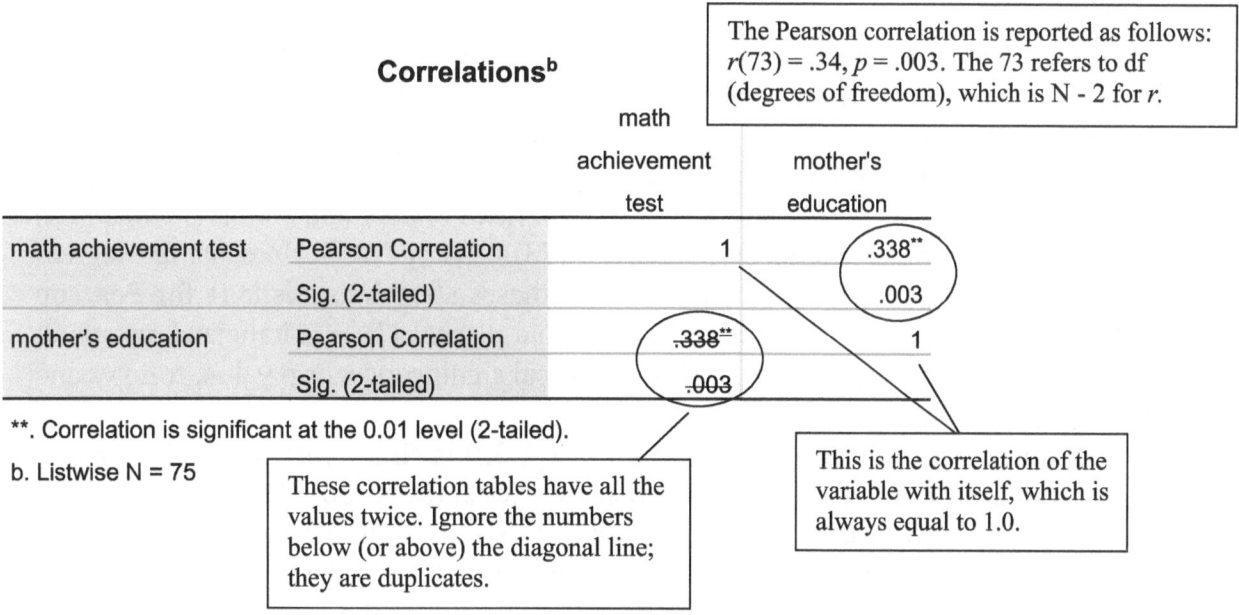

Correlations[b]

The Pearson correlation is reported as follows: $r(73) = .34, p = .003$. The 73 refers to df (degrees of freedom), which is N - 2 for *r*.

		math achievement test	mother's education
math achievement test	Pearson Correlation	1	.338**
	Sig. (2-tailed)		.003
mother's education	Pearson Correlation	.338**	1
	Sig. (2-tailed)	.003	

**. Correlation is significant at the 0.01 level (2-tailed).

b. Listwise N = 75

These correlation tables have all the values twice. Ignore the numbers below (or above) the diagonal line; they are duplicates.

This is the correlation of the variable with itself, which is always equal to 1.0.

Nonparametric Correlations

```
NONPAR CORR
  /VARIABLES=mathach maed
  /PRINT=SPEARMAN TWOTAIL NOSIG
  /MISSING=LISTWISE.
```

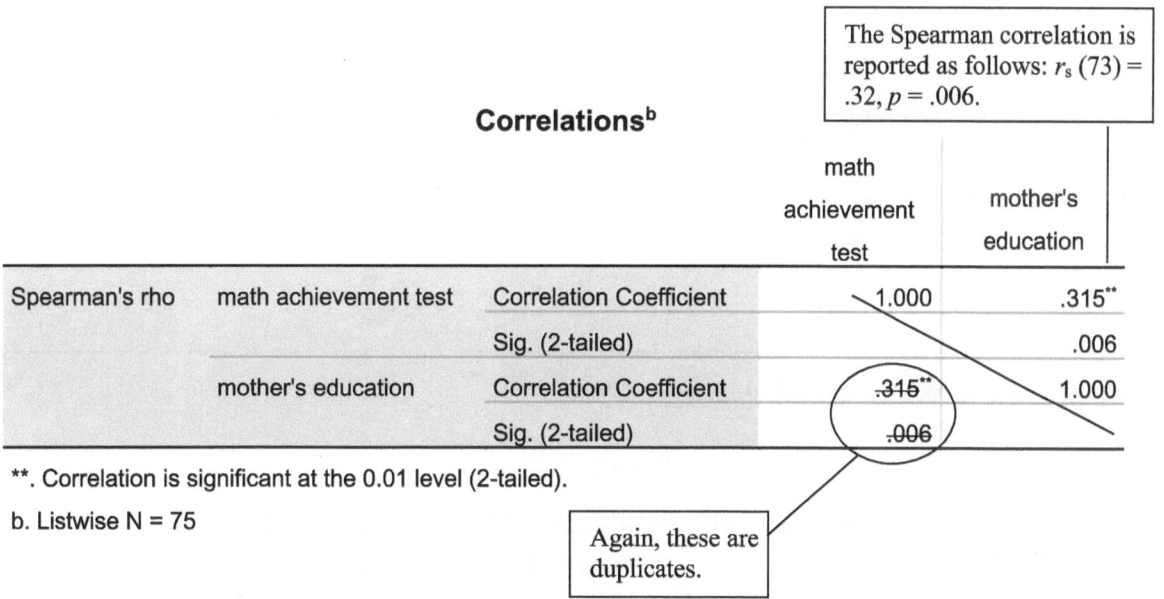

Correlations[b]

The Spearman correlation is reported as follows: $r_s (73) = .32, p = .006$.

			math achievement test	mother's education
Spearman's rho	math achievement test	Correlation Coefficient	1.000	.315**
		Sig. (2-tailed)		.006
	mother's education	Correlation Coefficient	.315**	1.000
		Sig. (2-tailed)	.006	

**. Correlation is significant at the 0.01 level (2-tailed).

b. Listwise N = 75

Again, these are duplicates.

Interpretation of Output 6.2

The first table provides **descriptive statistics** (mean, standard deviation, and *N*) for the variables to be correlated, in this case *math achievement* and *mother's education*. The two tables labeled **Correlations** are our primary focus. The information is displayed in matrix form by default, which unfortunately means that every number is presented twice. We have provided a call out box to help you.

The **Pearson Correlation** coefficient is .34; the significance level **(Sig.)** or p is .003 and the number of participants with both variables (*math achievement* and *mother's education*) is 75. In a report, this would usually be written as $r(73) = .34, p = .003$. Note that the degrees of freedom ($N - 2$ for correlations) is put in parentheses after the statistic (r for **Pearson** correlation), which is usually rounded to two decimal places. The r is italicized, as are all statistical symbols using English letters. The statistical significance, or p value, follows and is stated as $p = .003$.

The correlation value for **Spearman's rho** or r_s (.32) is slightly different from r, but usually, as in this case, it has a similar significance level ($p = .006$). The nonparametric Spearman correlation is based on ranking the scores (1st, 2nd, etc.) rather than using the actual raw scores. It should be used when the scores are ordinal data or when assumptions of the Pearson correlation (such as normality of the scores) are markedly violated. Note, you should *not* report both the Pearson and Spearman correlations; pick the one whose assumptions best fit the data. In this case, because *mother's education* was markedly skewed, Spearman would be the more appropriate choice. Problem 6.1 showed you a way to check the Pearson assumption that there is a linear relationship between the variables (i.e., that it is reasonable to use a straight line to describe the relationship).

It is usually best to choose two-tailed tests, as we did in Figure 6.6. We also chose to flag (put asterisks beside) the correlation coefficients that were statistically significant so that they could be identified quickly. The output also prints the exact significance level (p), which is more specific than just knowing it is statistically significant by seeing the asterisk. It is best in a thesis or paper table to report the exact p, but if space is tight, you can use asterisks with a footnote, as did Output 6.2.

Example of How to Write About Problem 6.2

Results

To investigate if there was a statistically significant association between mother's education and math achievement, a correlation was computed. Mother's education was skewed (skewness = 1.13), which violated the assumption of normality. Thus, the Spearman rho statistic was calculated, $r_s(73) = .32, p = .006$. The direction of the correlation was positive, which means that as mothers have increasing amounts of education, their children have increasingly higher math achievement test scores and vice versa. Using Cohen's (1988) guidelines, the effect size is medium for studies in this area. The r^2 indicates that approximately 10% of the variance in math achievement test scores can be predicted from mother's education.

Problem 6.3: Correlation Matrix for Several Variables

If you have more than two ordinal or normally distributed variables that you want to correlate, the program will produce a matrix showing the correlation of each selected variable with each of the others. You could create a matrix scatterplot to check the linear relationship for each pair of variables by clicking on **Graphs → Scatter/Dot** and selecting **Matrix Scatter** (see Figure 6.1). We did such a scatterplot (not shown) and found that the assumption of linearity was not markedly violated for this problem.

6.3. What are the associations among the four variables, *grades in h.s., math achievement test visualization test, and scholastic aptitude test—math?*

Now, compute **Pearson** correlations among all pairs of the following scale/normal variables: *grades in h.s., math achievement test, visualization test,* and *scholastic aptitude test—math.* Move all four into the **Variables** box (see Figure 6.6). Follow the procedures for Problem 6.2 outlined previously, except:

- Do not check Spearman (under **Correlation Coefficients**) but do check **Pearson**.
- Check **Show only the lower triangle** (otherwise, each correlation will appear twice).
- For **Options**, click **Means and standard deviations**, and **Exclude cases listwise**. The latter will only use participants who do not have missing data on any of these four variables.
- Click on **Continue**.
- Click on **Confidence interval**.
- Click on **Estimate confidence interval of bivariate correlation parameter**.
- Click on **Continue**.

This will produce Output 6.3. To see if you are doing the work right, compare your syntax and output to Output 6.3.

Output 6.3: Pearson Correlation Matrix

```
CORRELATIONS
  /VARIABLES=grades mathach visual satm
  /PRINT=TWOTAIL NOSIG LOWER
  /STATISTICS DESCRIPTIVES
  /MISSING=PAIRWISE.
```

Correlations

Descriptive Statistics

	Mean	Std. Deviation	N
grades in h.s.	5.68	1.570	75
math achievement test	12.5645	6.67031	75
visualization test	5.2433	3.91203	75
scholastic aptitude test - math	490.53	94.553	75

> Correlations of the other variables with grades in high school. Asterisks indicate significance at .01 (see footnote). For the other variables, you need to look in more than one column to see how they correlate with all other variables. All correlations are significant except the one between visualization and grades in high school.

Correlations[b]

		grades in h.s.	math achievement test	visualization test	scholastic aptitude test - math
grades in h.s.	Pearson Correlation	--			
math achievement test	Pearson Correlation	.504**	--		
	Sig. (2-tailed)	<.001			
visualization test	Pearson Correlation	.127	.423**	--	
	Sig. (2-tailed)	.279	<.001		
scholastic aptitude test - math	Pearson Correlation	.371**	.788**	.356**	--
	Sig. (2-tailed)	.001	<.001	.002	

**. Correlation is significant at the 0.01 level (2-tailed).
b. Listwise N=75

> The confidence interval tells you the "margin of error". The Lower and Upper limits of the confidence interval will be the same sign if the correlation is significant (meaning the confidence interval does not include 0 or no relationship). The true population correlation should be no smaller than the lower limit and no larger than the upper limit.

Confidence Intervals

	Pearson Correlation	Sig. (2-tailed)	95% Confidence Intervals (2-tailed)[a]	
			Lower	Upper
grades in h.s. - math achievement test	.504	<.001	.313	.656
grades in h.s. - visualization test	.127	.279	-.103	.344
grades in h.s. - scholastic aptitude test - math	.371	.001	.157	.551
math achievement test - visualization test	.423	<.001	.217	.593
math achievement test - scholastic aptitude test - math	.788	<.001	.683	.861
visualization test - scholastic aptitude test - math	.356	.002	.141	.540

a. Estimation is based on Fisher's r-to-z transformation.

> Even the lower limit of the confidence interval for the relation between grades in high school and math achievement is medium in size, suggesting this is likely a meaningful correlation.

Interpretation of Output 6.3

Notice that after the **Descriptive Statistics** table, there is a larger **Correlations** table that shows the **Pearson Correlation** coefficients and two-tailed significance **(Sig.)** levels. We have requested that these correlations NOT be given twice, so you need to look at more than one column to see all correlations with most variables. Note that significance is set at $p \leq .01$, rather than .05. This is to correct for the fact that we did multiple correlations. Because there are six correlations, the odds are increased that one could be statistically significant by chance. Thus, it is prudent to require a smaller value of p. The Bonferroni correction is a conservative approach designed to keep the significance level at .05 for the whole study. Using Bonferroni, you divide the usual significance level (.05) by the number of tests. In this case a $p < .008$ (.05/6) would be required for statistical significance. SPSS has rounded this to a more conventional level of .01. Another approach is simply to set alpha (the p value required for statistical significance) at a more conservative level, perhaps .01 instead of .05.

The Pearson correlations in this table are interpreted similarly to the one in Output 6.2. Note that if we had checked **Exclude cases pairwise** in Figure 6.7, the correlations would be the same, in this instance, because there were no missing data ($N = 75$) on any of the four variables. However, if some variables had missing data, the correlations would be at least somewhat different. Each correlation would be based on the cases that have no missing data <u>on those two variables</u>. One might use pairwise exclusion to include as many cases as possible in each correlation; however, the problem with this approach is that the correlations will include data from somewhat different individuals, making comparisons among correlations difficult. Multiple regression, which we will briefly discuss later in this chapter, uses listwise exclusion, including only the subjects with no missing data.

If you checked **One-Tailed Test of Significance** in Figure 6.6, the **Pearson Correlation** values would be the same as in Output 6.3, but the **Sig.** values would be half what they are here. For example, the Sig. for the correlation between *visualization test* and *grades in h.s.* would be .139 instead of .279. One-tailed tests are only used if you have a clear directional hypothesis (e.g., there will be a *positive* correlation between the variables). If one takes this approach, then all correlations in the direction opposite from that predicted must be ignored, even if they would otherwise be significant, so most researchers always use two-tailed tests.

Although the significance level for each correlation is important for determining whether or not results are likely to be due to chance, the actual value of the correlation is also important to note, as a measure of **effect size**. As you learned in Chapter 3, even if there is a relatively weak relationship between variables, you can obtain significance if you have a very large N, so in order to better understand whether the relationship is important and meaningful in the real world, it is useful to assess the actual size of the relationship. According to Cohen (1988), if the correlation is .30 or greater, it can be considered medium; if it is .50 or greater, it can be considered large. The confidence interval tells you what the size of the correlation is likely to be in the population. It is likely that the true correlation in the population is at least as large as the lower limit of the confidence interval but no larger than the upper limit of the confidence interval. So the relationship between *grades in high school* and *math achievement* is almost certainly at least moderate in size (lower limit .313) and is likely to be meaningful.

Example of How to Write About Problem 6.3

Results

Because each of the four achievement variables was normally distributed and the assumption of linearity was not markedly violated, Pearson correlations were computed to examine the intercorrelations of the variables. Table 6.3 shows that five of the six pairs of variables were significantly correlated. The strongest positive correlation, which would be considered a very large effect size according to Cohen (1988), was between the scholastic aptitude math test and math achievement test, $r(73) = .79, p < .001$. This means that students who had relatively high SAT math scores were very likely to have high math achievement test scores. Math achievement was also positively correlated with visualization test scores ($r = .42$) and grades in high school ($r = .50$); these are medium to large size effects or correlations according to Cohen (1988).

Table 6.3

Intercorrelations, Means, and Standard Deviations for Four Achievement Variables (N = 75)

Variable	1	2	3	4	M	SD
1. Visualization	—	.36**	.13	.42**	5.24	3.91
2. SAT math	—	—	.37**	.79**	490.53	94.55
3. Grades	—	—	—	.50**	5.68	1.57
4. Math ach.	—	—	—	—	12.56	6.67

*$p < .05$ **$p < .01$

Problem 6.4: Bivariate or Simple Linear Regression

As stated earlier, the Pearson correlation is the best choice for a statistic when you are interested in the simple association of two variables that both have normal or scale level measurement. Correlations do not indicate prediction of one variable from another; however, there are times when researchers wish to make such predictions. To do this, one needs to use regression (which is called bivariate (2-variable) or simple linear regression when there is just one predictor and one dependent variable). Assumptions and conditions for simple regression are similar to those for Pearson correlations; the variables should be approximately normally distributed and should have a linear relationship. However, for regression to be appropriate, one must be thinking about one variable as the predictor of the other one, for example if one is measured earlier than the other.

6.4. Can we predict *math achievement* from *grades in high school*?

To answer this question, a bivariate (two variables) regression is the best choice. Follow these commands, substituting your own variables if you are using your own dataset:

- **Analyze → Regression → Linear . . .**
- Highlight *math achievement*. Click the arrow to move it into the **Dependent** box, because we are thinking of it as the dependent variable.

- Highlight *grades in high school* and click on the arrow to move it into the **Independent(s)** box, because we are thinking of it as the predictor. The window should look like Figure 6.8.

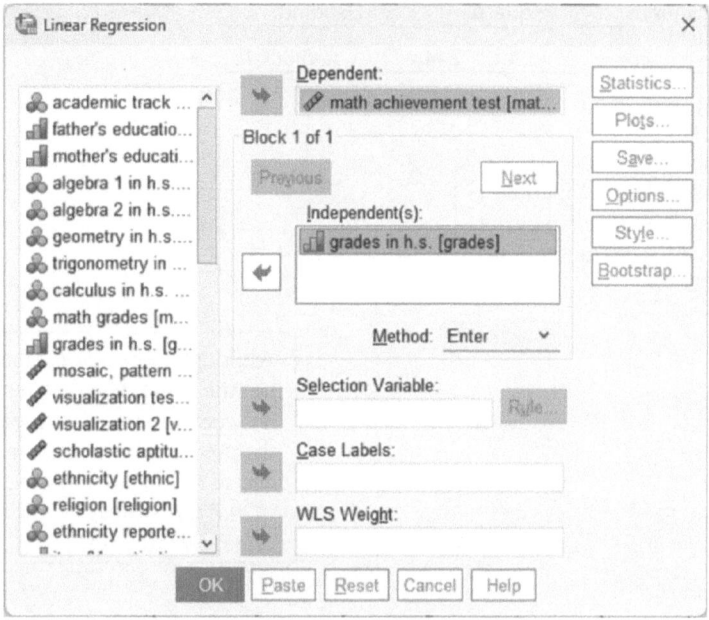

Figure 6.8 Linear regression.

- Click on **OK**.
- Compare your output with Output 6.4.

Output 6.4: Bivariate Regression

```
REGRESSION
  /MISSING LISTWISE
  /STATISTICS COEFF OUTS R ANOVA
  /CRITERIA=PIN(.05) POUT(.10)
  /NOORIGIN
  /DEPENDENT mathach
  /METHOD=ENTER grades.
```

Regression

Variables Entered/Removed[a]

Model	Variables Entered	Variables Removed	Method
1	grades in h.s.[b]		Enter

a. Dependent Variable: math achievement test

b. All requested variables entered.

Model Summary

Model	R	R Square	Adjusted R Square	Std. Error of the Estimate
1	.504[a]	.254	.244	5.80018

a. Predictors: (Constant), grades in h.s.

Note that the *R* in this table is a bivariate *r* between the *grades* and *math achievement*. This *R* is the same as the *r* in Output 7.3.

Although this table is labeled "ANOVA," it is not the usual ANOVA test of differences between groups. It presents a test of whether the predictor(s) significantly predict the DV. These results are reported as follows: $F(1,73) = 24.87, p < .001, r(73) = .50$. The *r* is an effect size measure.

ANOVA[a]

Model		Sum of Squares	df	Mean Square	F	Sig.
1	Regression	836.606	1	836.606	24.868	<.001[b]
	Residual	2455.875	73	33.642		
	Total	3292.481	74			

a. Dependent Variable: math achievement test

b. Predictors: (Constant), grades in h.s.

This is the regression coefficient (*b*), which is the slope of the best fit (regression) line. Note that it is not equal to the correlation coefficient. The **standardized** regression coefficient or Beta (.504) for simple regression is the correlation between those 2 variables.

Coefficients[a]

Model		Unstandardized Coefficients		Standardized Coefficients		
		B	Std. Error	Beta	t	Sig.
1	(Constant)	.397	2.530		.157	.876
	grades in h.s.	2.142	.430	.504	4.987	<.001

a. Dependent Variable: math achievement test

Interpretation of Output 6.4

The **Model Summary** table provides R, which in bivariate (two variable) regression is the same as r. The table also includes R^2 and an adjusted R^2 (.24), which indicates the proportion of the variance in math achievement (24%) that can be predicted from *grades in high school*. If this were a multiple regression, the R would be a correlation between the entire set of predictors and the dependent variable, and the adjusted R^2 is mainly adjusted for the number of predictors. The **Variables Entered/Removed** table indicates that we have entered one independent or predictor variable, *grades in h.s.* The **ANOVA** table provides the test of the statistical significance of the regression; it indicates that grades is a statistically significant predictor ($p < .001$) of *math achievement*.

In the fourth table, labeled **Coefficients**, the Unstandardized regression Coefficient in bivariate regression is simply the slope of the "best fit" regression line for the scatterplot showing the association between two variables. The Standardized regression Coefficient is equal to the correlation between those same two variables. (In Problem 6.5, multiple regression, we will see that when there is more than one predictor, the relation between correlation and regression becomes more complex, and there is more than one standardized regression coefficient.) The primary distinction between bivariate regression and bivariate correlation (e.g., Pearson) is that in regression, you want to predict one variable from another variable; whereas in correlation you simply want to know how those two variables are related.

The **Unstandardized Coefficients** give you a formula that you can use to predict the y scores (dependent variable) from the x scores (independent variable). Thus, if one did not have access to the real y score, this formula would tell one the best way of estimating an individual's y score based on that individual's x score. For example, if we want to predict *math achievement* for a similar group knowing only *grades in h.s.*, we could use the regression equation to estimate an individual's achievement score; predicted *math achievement* = 2.14 × (the person's *grades* score) + .40 the constant on intercept. Thus, if a student has mostly Bs (i.e., a code of 6) for their grades, their predicted *math achievement* score would be 13.24; *math achievement* = (2.14) (6) + .40.

One should be cautious in doing this, particularly with only one predictor, because we know (from the **Model Summary** table) that *grades in h.s.* only explains 24% of the variance in *math achievement*, so this would not yield a very accurate prediction. Another use of simple regression is to test a directional hypothesis: high *grades in h.s.* predict high *math achievement*. If one really thinks that this is the direction of the relationship (and not the reverse; i.e., *math achievement* leads to *grades in h.s.*), then regression is more appropriate than correlation.

How to Write About Output 6.4

Results

Simple regression was conducted to investigate how well grades in high school predict math achievement scores. The results were statistically significant, $F(1, 73) = 24.87$, $p < .001$. The identified equation to understand this relationship was math achievement = 2.14 × (grades in high school) + .40. The adjusted R^2 value was .244. This indicates that 24% of the variance in math achievement was explained by grades in high school. According to Cohen's (1988) guidelines, this is a large effect.

Problem 6.5: Multiple Regression

The purpose of multiple regression is similar to bivariate regression, but with more predictor variables. Multiple regression attempts to predict a normal (i.e., scale) dependent variable from a combination of several normally distributed and/or dichotomous independent/predictor variables. In this problem, we will see if *math achievement* can be predicted well from a combination of several of our other variables, *academic track*, *grades in high school*, and *mother's* and *father's education*. There are many different ways to analyze data with multiple regression. We will use an approach where we assume that all four of the predictor variables are important and that we want to see how these variables, taken together, predict the dependent variable. For this purpose, we will use the method the SPSS program calls **Enter** (usually called **simultaneous regression**), which tells the computer to consider all the variables at the same time. Our *IBM SPSS for Intermediate Statistics* book (Barrett et al., 2025) provides more examples and discussion of multiple regression assumptions, methods, and interpretation.

Assumptions and Conditions of Multiple Regression

There are many assumptions to consider, but we will only focus on the major ones that are easily tested. These include the following: the relationship between each of the predictor variables and the dependent variable is linear, the errors are normally distributed, and the variance of the residuals (difference between actual and predicted scores) is constant. A condition that can be problematic is **multicollinearity** (also called collinearity); it occurs when there are high intercorrelations among some set of the predictor variables. In other words, multicollinearity happens when overlapping or similar information in two or more predictors is predictive of the dependent variable.

6.5. How well can you predict *math achievement* from a combination of four variables: *grades in high school*, *father's and mother's education*, and *academic track*?

In this problem, the computer will enter or consider all the variables at the same time. We will ask which of these four predictors contribute significantly to the multiple correlation/regression when all are used together to predict *math achievement*.

Let's compute the regression for these variables. To do this, follow these steps:

- Click on the following: **Analyze → Regression → Linear ...** The Linear Regression window (Figure 6.9) should appear.

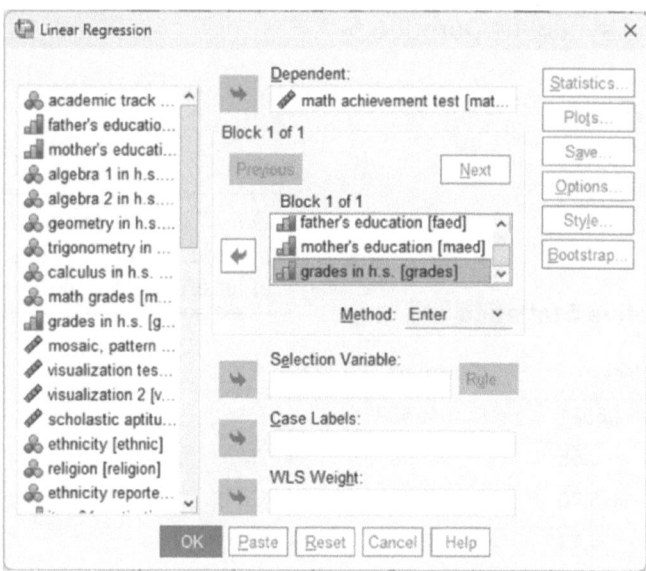

Figure 6.9 Linear regression.

- Select *math achievement test* and click it over to the **Dependent** box (dependent variable).
- Next select the variables *academic track, father's education, mother's education,* and *grades in h.s.,* and click them over to the **Independent(s)** box (independent variables).
- Under **Method**, be sure that **Enter** is selected.
- Click on **Statistics** at the top right corner of Figure 6.9 to get Figure 6.10.
- Make sure **Estimates** (under **Regression coefficients**) and **Model fit** are checked, and also check **Descriptives** and **Collinearity diagnostics** (see Figure 6.10).

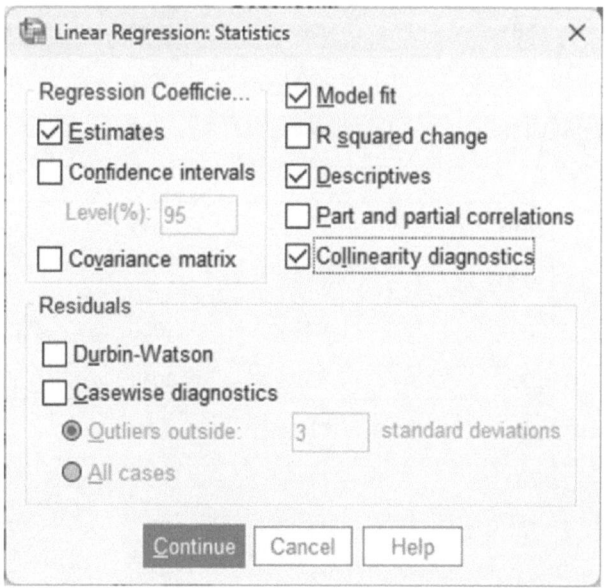

Figure 6.10 Linear regression: Statistics.

- Click on **Continue**.
- Click on **OK**.

Compare your output and syntax to Output 6.5.

Output 6.5: Multiple Regression

> Only 73 participants are
> included; 2 are missing
> at least one variable.

Descriptive Statistics

	Mean	Std. Deviation	N
math achievement test	12.6621	6.49659	73
academic track	.55	.501	73
grades in h.s.	5.70	1.552	73
father's education	4.73	2.830	73
mother's education	4.14	2.263	73

> Significance levels for correlations
> between predictors and achievement:
> all are significant or very close.

> All correlations (between pairs of
> the predictors) should be low to
> moderate, but this one is high.

Correlations

		math achievement test	academic track	father's education	mother's education	grades in h.s.
Pearson Correlation	math achievement test	1.000	-.274	.381	.345	.472
	academic track	-.274	1.000	-.265	-.202	.144
	father's education	.381	-.265	1.000	.681	.269
	mother's education	.345	-.202	.681	1.000	.190
	grades in h.s.	.472	.144	.269	.190	1.000
Sig. (1-tailed)	math achievement test	.	.010	<.001	.001	<.001
	academic track	.010	.	.012	.043	.112
	father's education	.000	.012	.	.000	.011
	mother's education	.001	.043	.000	.	.054
	grades in h.s.	.000	.112	.011	.054	.
N	math achievement test	73	73	73	73	73
	academic track	73	73	73	73	73
	father's education	73	73	73	73	73
	mother's education	73	73	73	73	73
	grades in h.s.	73	73	73	73	73

Variables Entered/Removed[a]

Model	Variables Entered	Variables Removed	Method
1	grades in h.s., academic track, mother's education, father's education[b]	.	Enter

a. Dependent Variable: math achievement test
b. All requested variables entered.

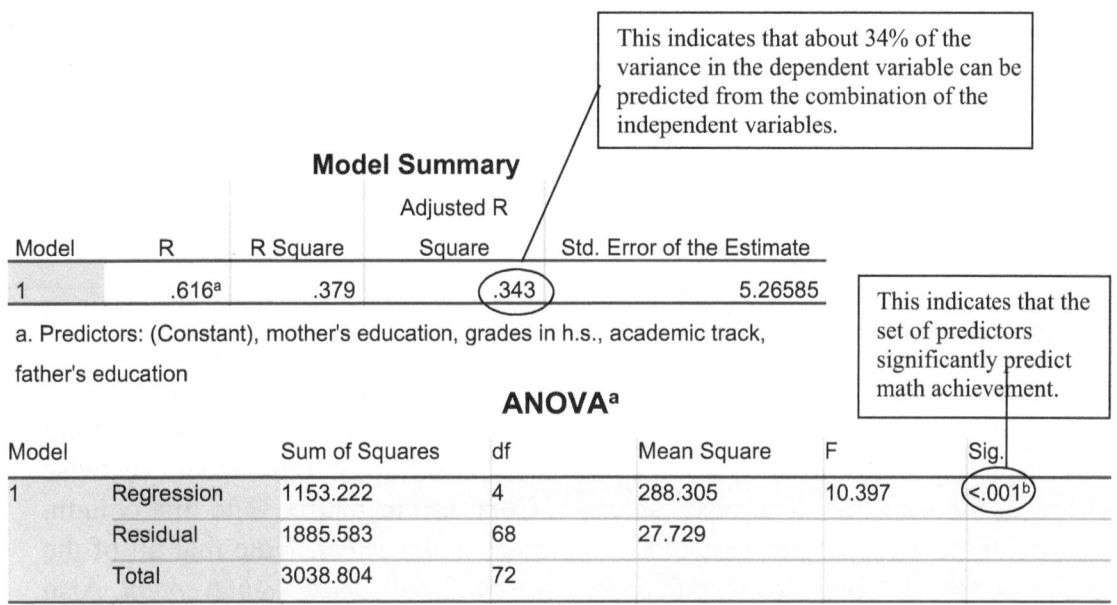

This indicates that about 34% of the variance in the dependent variable can be predicted from the combination of the independent variables.

Model Summary

Model	R	R Square	Adjusted R Square	Std. Error of the Estimate
1	.616[a]	.379	.343	5.26585

a. Predictors: (Constant), mother's education, grades in h.s., academic track, father's education

This indicates that the set of predictors significantly predict math achievement.

ANOVA[a]

Model		Sum of Squares	df	Mean Square	F	Sig.
1	Regression	1153.222	4	288.305	10.397	<.001[b]
	Residual	1885.583	68	27.729		
	Total	3038.804	72			

a. Dependent Variable: math achievement test

b. Predictors: (Constant), grades in h.s., academic track, mother's education, father's education

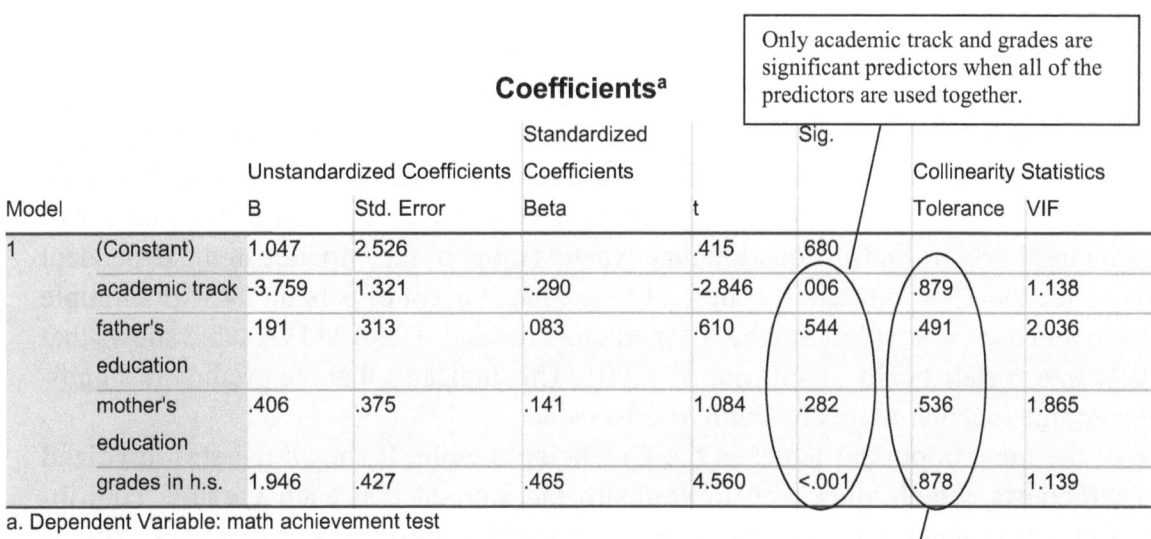

Only academic track and grades are significant predictors when all of the predictors are used together.

Coefficients[a]

Model		Unstandardized Coefficients B	Std. Error	Standardized Coefficients Beta	t	Sig.	Collinearity Statistics Tolerance	VIF
1	(Constant)	1.047	2.526		.415	.680		
	academic track	-3.759	1.321	-.290	-2.846	.006	.879	1.138
	father's education	.191	.313	.083	.610	.544	.491	2.036
	mother's education	.406	.375	.141	1.084	.282	.536	1.865
	grades in h.s.	1.946	.427	.465	4.560	<.001	.878	1.139

a. Dependent Variable: math achievement test

You want tolerances to be close to 1.0, indicating low collinearity. Here, both father's education and mother's education have low tolerances, indicating collinearity. Recall that these variables were highly correlated with each other. This may be one reason neither predicts math achievement significantly

Collinearity Diagnostics[a]

Model	Dimension	Eigenvalue	Condition Index	Variance Proportions				
				(Constant)	academic track	father's education	mother's education	grades in h.s.
1	1	4.214	1.000	.00	.01	.01	.01	.00
	2	.541	2.792	.00	.52	.04	.03	.00
	3	.134	5.616	.12	.45	.13	.15	.12
	4	.078	7.342	.01	.02	.79	.77	.01
	5	.034	11.177	.87	.00	.04	.04	.87

a. Dependent Variable: math achievement test

Interpretation of Output 6.5

This output begins with the usual **Descriptive Statistics** for all five variables in the first table. Note that the N is 73 because two participants are missing a score on one or more variables. Multiple regression uses only the participants who have complete data (listwise exclusion) for all the variables. The next table is a **Correlation** matrix. The first column shows the correlations of the other variables with *math achievement*. Note that all of the independent/predictor variables are significantly correlated with *math achievement*. Also notice that two of the predictor/independent variables are highly correlated with each other; *mother's* and *father's education* are correlated .68, which is not desirable. It might have been better to use only *mother's* (or *father's*) *education* or a combined *parents' education*. This is further supported by their low tolerances.

   The **Model Summary** table shows that the multiple correlation coefficient (R), using all the predictors simultaneously, is .62 and the **Adjusted R^2** is .34, meaning that 34% of the variance in math achievement can be predicted from the combination of *father's education, mother's education, grades in h.s.,* and *academic track*. Note that the adjusted R^2 is lower than the unadjusted R^2 (.38). This is, in part, related to the number of variables in the equation. Because several independent variables were used, a reduction of the number of variables might help us find an equation that explains more of the variance in the dependent variable, once the R^2 is adjusted. It is helpful to use the concept of parsimony with multiple regression and use the smallest number of predictors needed. The **ANOVA** table shows that $F = 10.40$ and is statistically significant, $p < .001$. This indicates that the predictors significantly combine together to predict *math achievement*.

   One of the most important tables is the **Coefficients** table. It shows the **standardized beta coefficients**, which allow you to compare the amount that each variable contributes to predicting *math achievement*, when all variables are used as predictors. This is because standardizing the coefficient puts all predictors on the same scale, so they can be compared. However, the unstandardized coefficients are better to use for predicting the dependent variable from the predictor scores. The *t* value and the **Sig.** opposite each independent variable indicates whether that variable is significantly contributing to the equation for predicting *math achievement*. Thus, *grades* and *academic track*, in this example, are the only variables that are significantly adding to the prediction. Each of them is significant <u>when the other three variables are already considered</u>, but the parent education variables add a little. It is important to note that all the variables are being considered together when these values are computed. Therefore, if you delete one of the predictors, even if it is not statistically significant, it can affect the levels of significance for other predictors. For example, if we deleted *father's education*, it is quite possible that *mother's education* would be a statistically significant predictor. The fact that both *father's education* and *mother's education* are correlated with *math achievement* and with each other makes this possibility more likely.

How to Write About Output 6.5

Results

Simultaneous multiple regression was conducted to investigate the best prediction of math achievement test scores from grades in high school, father's education, mother's education, and academic track. The means, standard deviations, and intercorrelations can be found in Table 6.4. This combination of variables significantly predicted math achievement, $F(4, 68) = 10.40$, $p < .001$. The beta coefficients are presented in Table 6.5. Note that higher grades and fast academic track significantly predict math achievement when all four variables are included. The adjusted R^2 value was .343. This indicates that 34% of the variance in math achievement was explained by the model. According to Cohen (1988), this is a large effect.

Table 6.4

Means, Standard Deviations, and Intercorrelations for Math Achievement and Predictor Variables
(N = 73)

Variable	M	SD	Grades in h.s.	Father's education	Mother's education	Academic track
Math achievement	12.66	6.50	.47**	.38**	.35**	−.27*
Predictor variables						
Grades in h.s.	5.70	1.55	—	.27*	.19	.14
Father's education	4.73	2.83		—	.68**	−.27*
Mother's education	4.14	2.26			—	−.20*
Academic track	.55	.50				—

*$p < .05$; **$p < .01$.

Table 6.5

Simultaneous Multiple Regression Analysis Summary for Grades in High School, Father's and Mother's Education, and Academic Track Predicting Math Achievement (N = 73)

Variable	B	SE B	β	t	p
Grades in h.s.	1.95	.43	.47	4.56	< .001
Father's education	.19	.31	.08	.61	.544
Mother's education	.41	.38	.14	1.08	.282
Academic track	3.86	1.32	−.29	−2.85	.006
Constant	1.05	2.53			

Note. $R^2 = .38$; $F(4, 68) = 10.40$, $p < .001$.

Interpretation Questions

6.1. Why would we graph scatterplots and regression lines?

6.2. In Output 6.2, (a) What do the correlation coefficients tell us? (b) What is r^2 for the Pearson correlation? What does it mean? (c) Compare the Pearson and Spearman correlations on both correlation size and significance level; (d) When should you use which type in this case? (d) Should you report r^2, the effect size, or both? Explain your answer.

6.3. In Output 6.3, how many of the Pearson correlation coefficients are significant? Write an interpretation of (a) one of the significant and (b) one of the nonsignificant correlations in Output 6.3. Include whether or not the correlation is significant, your decision about the null hypothesis, *and* a sentence or two describing the correlations in nontechnical terms. Include comments related to the sign and to the effect size.

6.4. Using Output 6.4, find the regression (B) coefficient or weight and the standardized regression (Beta) coefficient. (a) How do these compare to the correlation between the same variables in Output 6.3? (b) What does the regression (B) weight tell you? (c) Give an example of a research problem in which the Pearson correlation would be more appropriate than bivariate regression, and one in which bivariate regression would be more appropriate than Pearson correlation.

6.5. In Output 6.5, what do the standardized regression weights or coefficients tell you about the ability of the predictors to predict the dependent variable?

Extra SPSS Problems

Using the CollegeStudentData.sav set (see Appendix A on how to retrieve it), do the following problems. Print your outputs after typing your interpretations on them. Please circle the key parts of the output that you discuss.

6.1. What is the correlation between student's height and parent's height? Also produce a scatterplot. Interpret the results, including statistical significance, direction, and effect size.

6.2. Write a question that can be answered via correlational analysis with two approximately normal or scale variables. Run the appropriate statistics to answer the question. Interpret the results.

6.3. Make a correlation matrix using at least four appropriate variables. Identify, using the variable names, the two strongest and two weakest correlations. What were the *r* and *p* values for each correlation?

6.4. Is there a combination of hours of TV watching, hours of studying, and hours of work that predicts current GPA?

7 Comparing Groups with *t* Tests, Analysis of Variance (ANOVA), and Similar Nonparametric Tests

In this chapter, we will discuss different statistics for comparing groups. First, in Problem 7.1, we will use a **one-sample *t* test** to compare one group or sample to a hypothesized population mean. Then, in Problems 7.2–7.5, we will examine two parametric and two nonparametric/ordinal statistics that compare two groups of participants. Problem 7.2 compares two independent groups (between-groups design), students in fast-track and students in regular track academics, using the **independent samples *t* test**. Problem 7.3 uses the **Mann-Whitney** nonparametric test, which is similar to the independent *t* test but is used when the assumptions of the independent *t* test are not met. Problem 7.4 is a within-subjects design that uses a **paired samples *t*** to compare the average or mean levels of education of students' mothers and fathers. Problem 7.5 shows how to use the nonparametric **Wilcoxon** test for a within-subjects design.

The top right side of Table 7.1 distinguishes between **between-groups** and **within-subjects designs**. This helps determine the specific statistic to use. The other determinant of which statistic to use has to do with statistical assumptions. If the assumptions are not markedly violated, you can use a parametric test, such as a *t* test or ANOVA. If the assumptions are markedly violated, one can use a nonparametric test, which does not have the same assumptions, as indicated by the bottom row left side of Table 7.1. Another alternative is to transform the variable so that it meets the assumptions. That is beyond the scope of this book, but is covered in Barrett et al. (2025).

Later in this chapter, you will learn about **ANOVA** (*F*), which can be used to look at differences between the means of <u>two or more groups</u>. You might ask, why is there the need for a *t* test when one-way ANOVA can be used to compare *two* groups as well as three or more groups? Because $F = t^2$, both statistics provide the same results. Thus, you could do an ANOVA even if there are only two groups. However, we recommend doing a *t* test if you have only two groups, especially under at least two circumstances: First, *t* tests can be either one-tailed or two-tailed, while one cannot have one-tailed ANOVAs. Therefore, if you have a clear directional hypothesis that predicts which group will have the higher mean, you may want to use a *t* test rather than one-way ANOVA when comparing two groups. Second, as mentioned in an earlier chapter, one of the assumptions of both the *t* test and ANOVA is that the groups have equal variances. Violating this assumption can cause the test to be inaccurate, especially if the number of participants in the groups varies. The SPSS *t* test program provides an adjustment to deal with the problem of unequal variances, whereas the remedy for such problems in ANOVA may be less

DOI: 10.4324/9781003355908-7

satisfactory; so if you have only two groups and unequal variances, it is better to do a *t* test. Finally, if you do a *t* test using SPSS, SPSS can calculate the best measure of effect size for a difference between two groups—the *d* (see chapter 3). Not surprisingly, it is more customary to use a *t* test if one is comparing only two groups. You *must* use ANOVA if you want to compare three or more groups.

You will learn how to compute two types of analysis of variance (ANOVA) and a similar non-parametric statistic. In Problem 7.6, we will use the **one-way or single factor ANOVA** to compare three levels of father's education on several dependent variables (e.g., math achievement). If the ANOVA is statistically significant, you will know that there is a difference somewhere, but you will not know which pairs of means were significantly different. In Problem 7.7, we show you when and how to do appropriate **post hoc tests** to see which pairs of means were different. In Problem 7.8, you will compute the **Kruskal-Wallis (K-W)** test, a nonparametric test similar to one-way ANOVA. In Problem 7.9, we will introduce you to **two-way or factorial ANOVA**, which is not shown in Table 7.1. Table 7.2 shows how to interpret measures of effect size that are used in this chapter.

Table 7.1

Selection of an Appropriate Inferential Statistic for Basic, Two Variable Difference Questions or Hypotheses

Scale of Measurement of **Dependent Variable**	Appropriate **Descriptive** Statistics to report	One Factor or Independent Variable With **Two Levels** or **Categories**/Groups/Samples		One Independent Variable **Three or More Levels** or Groups	
		Independent Samples or Groups **(Between)**	Repeated Measures or Related Samples **(Within)**	Independent Samples or Groups **(Between)**	Repeated Measures or Related Samples **(Within)**
Dependent Variable Approximates **Normal/Scale** Data and Assumptions Not Markedly Violated	Mean, Standard Deviation	INDEPENDENT SAMPLES *t* TEST	PAIRED SAMPLES *t* TEST	ONE-WAY ANOVA	GLM REPEATED MEASURES ANOVA I.B.[a]
Dependent Variables Clearly **Ordinal** or Parametric Assumptions Markedly Violated	Median (or Mean Rank), Interquartile Range	MANN-WHITNEY	WILCOXON	KRUSKAL-WALLIS	FRIEDMAN I.B.[a]

[a] This complex statistic is discussed in more detail in our companion book, Barrett et al. (2025), *IBM SPSS for Intermediate Statistics* (6th ed.).

Table 7.2.

Interpretation of the Strength of a Relationship (Effect Sizes) for Statistics Which Compare Groups

General Interpretation of the Strength of a Relationship	The d Family[a]	The r Family [b]					
	d	r^2	r, rho, ϕ	R^2	R	η^2	η (eta)[d]
Much larger than typical	$\geq |1.00|$[c, e]	.49	$\geq |.70|$	.49+	$|.70|+$	.21	$|.45|+$
Large or larger than typical	$|.80|$	.25	$|.50|$	.26	$|.51|$	.14	$|.37|$
Medium or typical	$|.50|$	.09	$|.30|$	.13	$|.36|$	.06	$|.24|$
Small or smaller than typical	$|.20|$	.01	$|.10|$	.02	$|.14|$	.01	$|.10|$

[a] *d* values can vary from 0.0 to + or − infinity, but *d* greater than 1.0 is relatively uncommon.

[b] *r* family values can vary from 0.0 to + or − 1.0, but except for reliability (i.e., same concept measured twice), *r* is rarely above .70. In fact, some of these statistics (e.g., phi) have a restricted range in certain cases; that is, the maximum phi is less than 1.0.

[c] We interpret the numbers in this table as a range of values. For example, a *d* greater than .90 (or less than −.90) would be described as "much larger than typical," a *d* between, say, .70 and .90 would be called "larger than typical," and *d* between, say, .60 and .70 would be "typical to larger than typical." We interpret the other three columns similarly.

[d] Partial etas are multivariate tests equivalent to *R*. Use *R* column.

[e] Note. | | indicates absolute value of the coefficient. The absolute magnitude of the coefficient, rather than its sign, is the information that is relevant to effect size. *R* and η usually are calculated by taking the square root of a squared value, so that the sign usually is positive.

- Retrieve **hsbdataNEW** from your data file. See Appendix A for how to retrieve the dataset if you need help.

Problem 7.1: One-Sample *t* Test

Sometimes you want to compare the mean of a sample with a hypothesized population mean to see if your sample is significantly different. For example, the scholastic aptitude test was originally standardized so that the mean was 500 and the standard deviation was 100. In our modified HSB dataset, we made up mock *scholastic aptitude test—math* (*SAT Math*) scores for each student. You may remember from Chapter 2 that the mean *SAT Math* score for our sample was 490.53. Is this significantly different from 500?

7.1. Is the mean *SAT Math* score in the modified HSB dataset statistically significantly different from the presumed population mean of 500?

Assumptions of the One-Sample t Test

1. The dependent variable is normally distributed within the population.
2. The data are independent (scores of one participant are not dependent on scores of the others; participants are independent of one another).

To compute the one-sample *t* test, use the following commands:

- **Analyze → Compare Means and proportions → One-Sample T Test . . .**
- Move *scholastic aptitude test—math* to the **Test Variable(s)** box.
- Type 500 in the **Test Value** box (the test value is the score that you want to compare to your sample mean). Make sure that the **Estimate effect sizes** box is checked.
- Your window should look like Figure 7.1.
- Click **OK**.

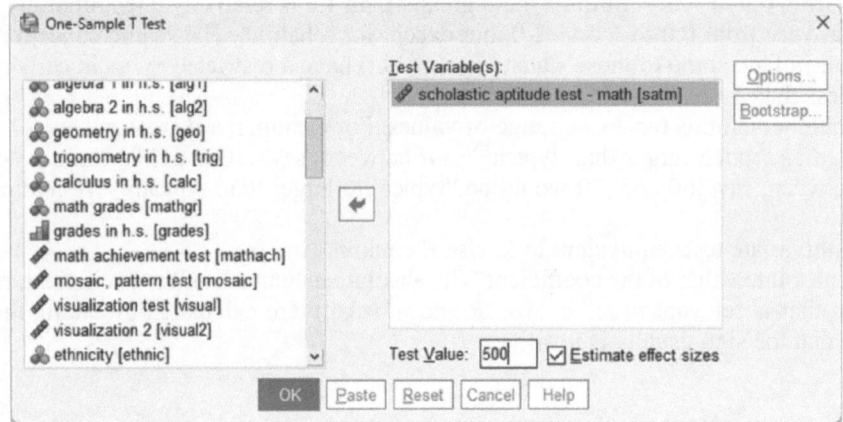

Figure 7.1 One-sample t test.

Compare your output to Output 7.1.

Output 7.1: One-Sample **t** *Test*

```
T-TEST
  /TESTVAL=500
  /MISSING=ANALYSIS
  /VARIABLES=satm
  /ES DISPLAY(TRUE)
  /CRITERIA=CI(.95).
```

This is the average *SAT math* score in the HSB sample.

One-Sample Statistics

	N	Mean	Std. Deviation	Std. Error Mean
scholastic aptitude test - math	75	490.53	94.553	10.918

This *t* test compares the sample mean of 490.53 with the test value of 500.

One-Sample Test

The *t* test is not significant, even if you use a one-tailed (one-sided) test.

Test Value = 500

	t	df	Significance		Mean Difference	95% Confidence Interval of the Difference	
			One-Sided p	Two-Sided p		Lower	Upper
scholastic aptitude test - math	-.867	74	.194	.389	-9.467	-31.22	12.29

Cohen's d is only -.10, which is a very small effect size. In fact, the upper and lower limits of the confidence interval are different signs, indicating that the true effect size might be zero.

One-Sample Effect Sizes

		Standardizer[a]	Point Estimate	95% Confidence Interval	
				Lower	Upper
scholastic aptitude test - math	Cohen's d	94.553	-.100	-.327	.127
	Hedges' correction	95.525	-.099	-.323	.126

a. The denominator used in estimating the effect sizes.

Cohen's d uses the sample standard deviation.

Hedges' correction uses the sample standard deviation, plus a correction factor.

Interpretation of Output 7.1

The **One-Sample Statistics** table provides basic descriptive statistics for the variable under consideration. The **Mean** *SAT Math* for the students in the sample was compared to the hypothesized population mean, displayed as the **Test Value** in the **One-Sample Test** table. On the bottom line of this table are the *t* value, *df*, and the two-tailed Sig. (*p*) value, which are encircled. Note that $p = .389$, so we can say that the sample mean ($M = 491$) is not significantly different from the population mean of 500. The table also provides the difference ($M = -9.47$) between the sample and population means and the 95% **Confidence Interval**. The difference between the sample and the population mean is likely to be between $+12.29$ and -31.22 points. Notice that this range includes the value of zero, so it is possible that there is no difference. Thus, the difference is not statistically significant.

Problem 7.2: Independent Samples *t* Test

When investigating the difference between two unrelated or independent groups (in this case fast track and regular track students) on an approximately normal dependent variable, it is appropriate to choose an independent samples *t* test if the following assumptions are not markedly violated.

Assumptions of the Independent Samples t Test

1. The variances (standard deviation squared) of the dependent variable in the two populations are equal.
2. The dependent variable is normally distributed within each population.
3. The data are independent (scores of one participant are not related systematically to scores of the others).

SPSS will automatically test Assumption 1 with the **Levene's test** for equality of variances. Assumption 2 could be tested, as we did in Chapter 2, Problem 2.6, with the **Explore** command, to see whether the dependent variables are at least approximately normally distributed for each academic track. Because the *t* test is quite robust to violations of this assumption, especially if the data for both groups are skewed in the same direction, we won't test it here. Assumption 3 probably is met because the academic tracks are not matched or related pairs, and there is no reason to believe that one person's score might have influenced another person's. This assumption is best addressed during design and data collection. In addition to ensuring that the data meet these assumptions, the researcher should try to ensure that groups or samples are of similar size, as the assumption of homogeneity of variance is most important and more likely to be violated if samples differ markedly in size.

7.2. Do fast track and regular track students differ in regard to their average *math achievement* scores, *grades in high school*, and *visualization test* scores?

One feature of this SPSS program is that it can do several *t* tests in a single output if they have the same independent or grouping variable (e.g., academic track). In this problem, we compute three separate *t* tests, one each for *math achievement, grades in high school*, and *visualization test* scores; in each, fast track students are compared to regular track students.

   With more than one dependent variable, one could have chosen to use MANOVA (see Intermediate book), especially if these variables were conceptually related and correlated with each other. MANOVA would enable us to see how a linear *combination* of these three variables was different for students in the fast track than for students in the regular track. We will not demonstrate MANOVA in this book, but see Barrett et al. (2025) *IBM SPSS for Intermediate Statistics* (6th ed.) for how to compute and interpret MANOVA.

For the *t* tests, follow these commands:

- Click on **Analyze → Compare means and proportions → Independent-Samples T Test . . .**
- Move *math achievement*, *grades in h.s.*, and *visualization test* to the **Test** (dependent) **Variable(s)** box and move *academic track* to the **Grouping** (independent) **Variable:** box (see Figure 7.2).

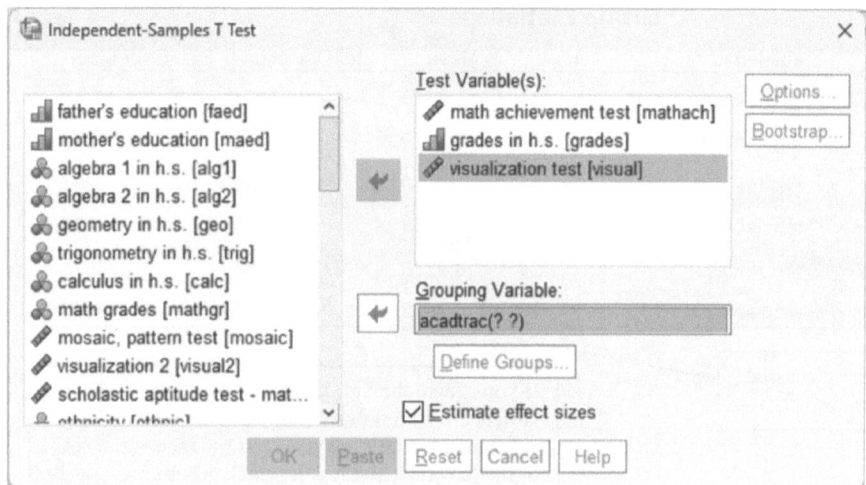

Figure 7.2 Independent-samples t test.

- Next click on **Define Groups . . .** in Figure 7.2 to get Figure 7.3.
- Type **0** (the number we assigned to fast track) in the **Group 1** box and **1** (the number we assigned for regular track) in the **Group 2** box (see Figure 7.3). This will enable us to compare fast track and regular track students on each of the three dependent variables.

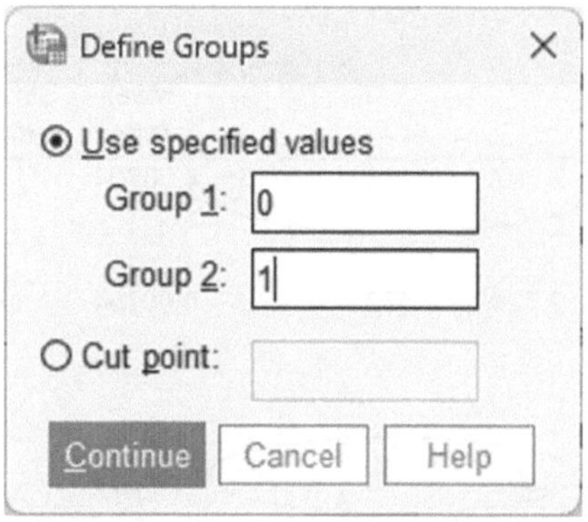

Figure 7.3 Define groups.

- Click on **Continue**, make sure the box by **Estimate effect sizes** is checked, then click on **OK**. Compare your output to Output 7.2.

Output 7.2: Independent Samples t Test

```
T-TEST GROUPS=acadtrac(0 1)
  /MISSING=ANALYSIS
  /VARIABLES=mathach visual grades
  /ES DISPLAY(TRUE)
  /CRITERIA=CI(.95).
```

T-Test

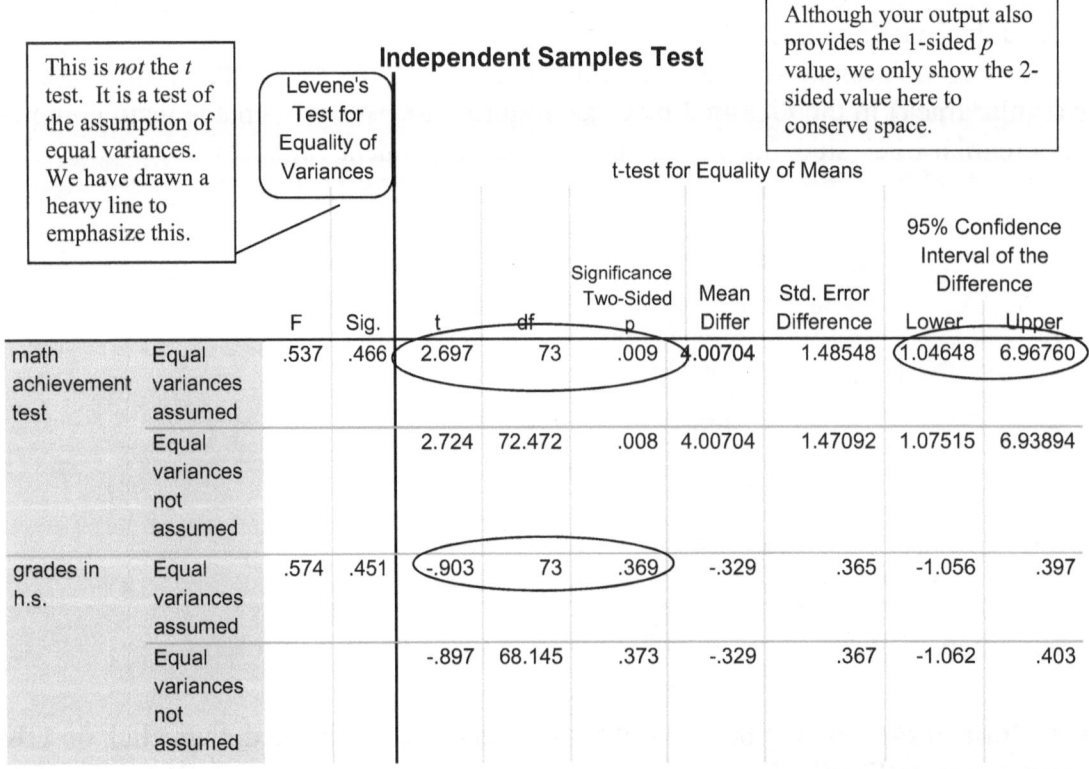

Each circle contains a pair of means to be compared.

Group Statistics

	academic track	N	Mean	Std. Deviation	Std. Error Mean
math achievement test	fast track	34	14.7550	6.03154	1.03440
	regular track	41	10.7479	6.69612	1.04576
grades in h.s.	fast track	34	5.50	1.638	.281
	regular track	41	5.83	1.515	.237
visualization test	fast track	34	6.4265	4.47067	.76671
	regular track	41	4.2622	3.10592	.48506

Circled numbers are discussed in the Interpretation box.

If you square the Std. Deviation you will get the variance. The left two columns of the next table show the Levene's test used to test if the variances of the fast track and regular track groups are significantly different on the visualization test variables.

This is *not* the *t* test. It is a test of the assumption of equal variances. We have drawn a heavy line to emphasize this.

Although your output also provides the 1-sided *p* value, we only show the 2-sided value here to conserve space.

Independent Samples Test

		Levene's Test for Equality of Variances		t-test for Equality of Means					95% Confidence Interval of the Difference	
		F	Sig.	t	df	Significance Two-Sided p	Mean Differ	Std. Error Difference	Lower	Upper
math achievement test	Equal variances assumed	.537	.466	2.697	73	.009	4.00704	1.48548	1.04648	6.96760
	Equal variances not assumed			2.724	72.472	.008	4.00704	1.47092	1.07515	6.93894
grades in h.s.	Equal variances assumed	.574	.451	-.903	73	.369	-.329	.365	-1.056	.397
	Equal variances not assumed			-.897	68.145	.373	-.329	.367	-1.062	.403

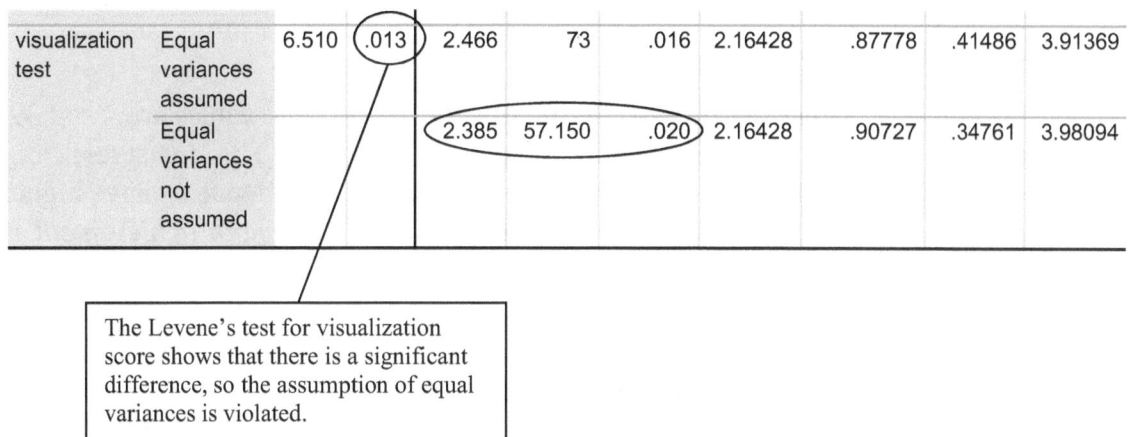

visualization test	Equal variances assumed	6.510	.013	2.466	73	.016	2.16428	.87778	.41486	3.91369
	Equal variances not assumed			2.385	57.150	.020	2.16428	.90727	.34761	3.98094

The Levene's test for visualization score shows that there is a significant difference, so the assumption of equal variances is violated.

Interpretation of Output 7.2

The first table, **Group Statistics**, shows descriptive statistics for the two groups (fast track and regular track students) separately. Note that the means within each of the three pairs look somewhat different. This might be due to chance, so we will check the *t* tests in the next table.

The second table, **Independent Samples Test**, provides two statistical tests. The left two columns of numbers are the **Levene's test** for the assumption that the variances of the two groups are equal. This is *not* the *t* test; it only assesses an assumption! If this *F* test is significant (as in the case of *visualization*), the variances are significantly different for the two groups, the assumption is *violated*, and one uses the **Equal variances *not* assumed** line for the *t* test and related statistics. However, if Levene's *F* is not statistically significant (Sig. > .05), as is true for *grades in high school* and *math achievement*, then variances are not significantly different statistically and the assumption of equal variances is not violated. In that case, the **Equal variances assumed** line is used. The appropriate lines to use are circled in the output.

Independent Samples Effect Sizes

		Standardizer[a]	Point Estimate	95% Confidence Interval Lower	Upper
math achievement test	Cohen's d	6.40424	.626	.158	1.089
	Hedges' correction	6.47099	.619	.156	1.078
	Glass's delta	6.69612	.598	.122	1.068
grades in h.s.	Cohen's d	1.572	-.210	-.665	.247
	Hedges' correction	1.588	-.207	-.658	.245
	Glass's delta	1.515	-.217	-.673	.241
visualization test	Cohen's d	3.78431	.572	.106	1.034
	Hedges' correction	3.82376	.566	.105	1.023
	Glass's delta	3.10592	.697	.213	1.172

Cohen's *d* is the first number in the **Point Estimate** column for each DV.

a. The denominator used in estimating the effect sizes.
Cohen's d uses the pooled standard deviation.
Hedges' correction uses the pooled standard deviation, plus a correction factor.
Glass's delta uses the sample standard deviation of the control group.

Interpretation of Output 7.2 continued

Thus, in the section titled **t-test for equality of means**, the appropriate values are: $t = 2.39$, degrees of freedom (df) = 57.15, and $p = .020$ for *visualization test*. This t is statistically significant, so based on examining the means, we can say that fast track students have higher *visualization* scores than regular track students. Likewise, the t for *math achievement* is statistically significant because $p = .009$. Thus, fast track students have higher means. However, for *grades in high school* the t is not statistically significant ($p = .369$), so we conclude that there is not enough evidence to say that there is a systematic difference between students in the fast track and students in the regular track on grades.

The **95% Confidence Interval of the Difference** is shown in the two right-hand columns of the **Independent Samples** table. If we constructed an infinite number of studies using the same conditions, and computed a 95% confidence interval for each study, 95% of the intervals would contain the true population difference between means. For *math achievement* the 95% confidence interval is between 1.05 points and 6.97 points. Note that if the **Upper** and **Lower** bounds have the same sign (either + and + or − and −), we know that the difference is statistically significant because this means that the null finding of zero difference lies *outside* of the confidence interval. On the other hand, if zero lies between the upper and lower limits, there could be no difference, as is the case for *grades in h.s.* The lower limit of the confidence interval on *math achievement* tells us that the difference between fast track and regular track could be as small as 1.05 points out of 25, which is the maximum possible score.

The final table shows **effect sizes** for the t tests. In addition to Cohen's *d,* Hedge's *g* and Glass' delta are provided. Hedge's *g* corrects for bias that may exist in Cohen's *d* when sample sizes are small. Glass's delta uses the distribution of the control group for experimental studies. We will report and discuss Cohen's *d* for three reasons: (1) the issues of bias in statistical estimation are beyond the scope of this class, (2) Cohen's *d* is more widely used and understood, and (3) it is better to use *d* in meta-analyses, which makes it more desirable to report. For *math achievement*, Cohen's *d* is .626, which is, according to Cohen (1988) and Table 7.2, a medium-sized "effect." There is also a medium "effect" on *visualization,* but a small "effect" on *grades.* Because you need means and standard deviations to fully understand the group difference and effect size, you should include a table with means and standard deviations in your results section for a full interpretation of *t* tests.

How to Write About Output 7.2

Results

Table 7.3 shows that fast track students were statistically significantly different from regular track students on *math achievement,* $t(73) = 2.70, p = .009$. Inspection of the two group means indicates that the average math achievement score for regular track students ($M = 10.75$) is significantly lower than the score ($M = 14.76$) for fast track students. The difference between the means is 4.01 points on a 25-point test. The effect size d is .63, which is a medium size for effects in the behavioral sciences. Fast track students did not differ from regular track students on grades in high school, $t(73) = -90, p = .369$, but fast track students did score higher on the visualization test, $t(57.2) = 2.39, p = .020$, which was statistically significant. The effect size, d, was .57, a medium size.

> It is important to include all this information either in a table or in the text.

Table 7.3

Comparison of Students in the Fast Track and Students in the Regular Track on a Math Achievement Test, Grades, and a Visualization Test (n = 34 fast track students and 41 regular track students)

Variable	M	SD	t	df	p	D
Math achievement			2.70	73	.009	.63
Fast track	14.76	6.03				
Regular track	10.75	6.70				
Grades			−.90	73	.369	−.21
Fast track	5.50	1.64				
Regular track	5.83	1.52				
Visualization			2.39[a]	57.2[a]	.020	.57
Fast track	6.43	4.47				
Regular track	4.26	3.11				

[a]The *t* and *df* were adjusted because variances were not equal.

Problem 7.3: The Nonparametric Mann-Whitney *U* Test

What should you do if the *t* test assumptions are markedly violated (e.g., what if the dependent variable data are grossly skewed, otherwise non-normally distributed, or are ordinal)? One answer is to run the appropriate nonparametric statistic, which in this case is called the Mann-Whitney (M-W) *U* test. The M-W is used with a between-groups design with two levels of the independent variable.

7.3. Do students on the fast track and regular track differ on *visualization, math achievement,* and *grades*?

For this problem, we will assume that the scores for the three dependent variables were ordinal level data or that other assumptions of the *t* test were violated but that the assumptions of the Mann-Whitney test were met.

Assumptions of the Mann-Whitney Test

1. It is assumed that scores are ordered from low to high on the dependent variable, before ranking.
2. The data are independent (scores of one participant are not dependent on scores of the others).

- Click on **Analyze → Nonparametric Tests → Independent Samples**
- Click on the **Fields** tab.
- Hold down the ctrl button and click to highlight all variables in the **Test Fields** box other than *visualization test, math achievement,* and *grades in h.s.* (the dependent variables) and then move them out of the box using the curved arrow to the left of the box.

- Next, click on *academic track* (the independent variable) and move it over to the **Group** box.
- Your window should look like Figure 7.4.

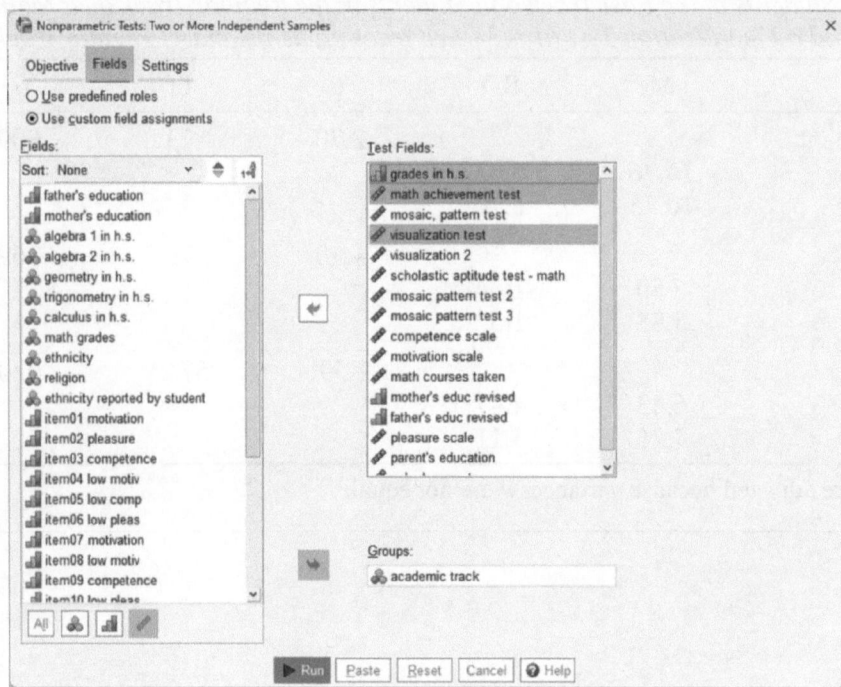

Figure 7.4 Nonparametric Tests: Two or More Independent Samples

- Click on **Run**

Compare your syntax and output to Output 7.3 to check your work.

Output 7.3: Nonparametric Test: Mann-Whitney U

```
*Nonparametric Tests: Independent Samples.
NPTESTS
  /INDEPENDENT TEST (grades mathach visual) GROUP (acadtrac)
  /MISSING SCOPE=ANALYSIS USERMISSING=EXCLUDE
  /CRITERIA ALPHA=0.05 CILEVEL=95.
```

Independent-Samples Mann-Whitney U Test

math achievement test across academic track
Independent-Samples Mann-Whitney U Test

Summary

Total N	75
Mann-Whitney U	455.500
Wilcoxon W	1316.500
Test Statistic	455.500
Standard Error	93.795
Standardized Test Statistic	-2.575
Asymptotic Sig.(2-sided test)	.010

> The **Mann-Whitney U** statistic is 455.5. The *p* value (Sig.) is **asymptotic**, meaning it is an estimate, rather than an exact number.

> Note *z* (**Standardized Test Statistic**) for each dependent variable. You will need them to calculate effect size, and when you write about your results.

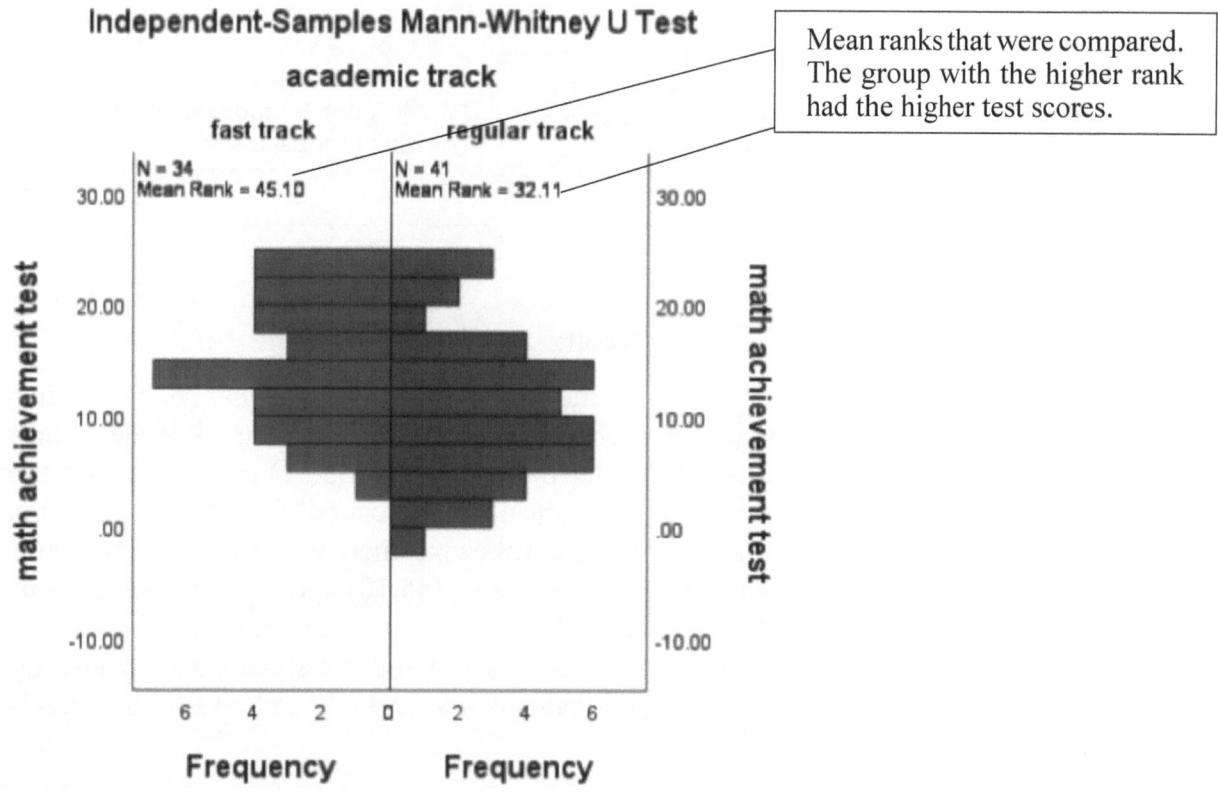

> Mean ranks that were compared. The group with the higher rank had the higher test scores.

Illustration 7.1 Independent Samples Mann-Whitney U Test Academic Track

Interpretation of Output 7.3

The first table provides the **Mann-Whitney *U*, *z* score (Standardized Test Statistic)**, and the **Sig.** (significance) level or *p* value for the **U test** for *math achievement*. To save space, we only show you the information for one DV, but the others are similar. Asymptotic significance refers to the fact that the significance levels are not exact. Note that the mean ranks of the academic tracks differ significantly on *visualization test* and *math achievement* but not on *grades in high school*, as was the case for the similar *t* tests in Problem 7.2. The Mann-Whitney test is only slightly less powerful than the *t* test, so it is a good alternative if the assumptions of the *t* test are violated, as was the case with *visualization test*. Note that you <u>would not report</u> both *t* tests and Mann-Whitney tests for the same variables because they provide very similar information.

Although an effect size measure is not provided in the output, it is easy to compute an *r* from the *z* provided in the **Test Statistics** table, using the conversion formula, $r = \frac{z}{\sqrt{N}}$. For *math achievement*, the effect size is $-2.75/\sqrt{75} = -2.75/8.66 = -.318$. You can see from Table 7.2 that this is a medium effect size. The *r* effect sizes are somewhat smaller than for the corresponding *t* tests in Output 7.2.

When conducting a nonparametric statistic, it is important to include why you chose to not use a parametric statistic.

Be sure to include the mean ranks for each level.

How to Write About Output 7.3

Results

Because the dependent variables were ordinal and the variances were unequal, Mann-Whitney *U* tests were performed to compare the academic tracks. The 34 fast track students have higher mean ranks (43.65) than the 41 regular track students (33.32) on the visualization test, $U = 505$, $p = .04$, $r = -.24$, which was a statistically significant difference and, according to Cohen (1988), is a small to medium effect size. Likewise, there was a statistically significant difference, in the mean ranks of fast track students (45.10) and regular track students (32.11) on math achievement, $U = 455.5$, $p = .01$, $r = -.32$, which is considered a medium effect size. However, fast track and regular track students did not show a statistically significant difference on grades in high school. Mean ranks were 35.78 and 39.84, respectively, $U = 621.5$, $p = .41$, $r = -.09$.

Problem 7.4: Paired Samples *t* Test

In this problem, you will compare the average scores of each HSB student's father's and mother's scores on the same measure, namely, their educational level. Because *father's* and *mother's education* are not independent of each other, the paired *t* test is the appropriate test to perform. The paired samples *t* test can also be 1   Q`1 used when the two scores are used for interrater reliability, such as Problem 4.2 in Chapter 4. Other examples would be in a longitudinal study or in a single group quasi-experimental study in which the same assessment is used as the pretest, before the intervention, and as the posttest, after the intervention.

Assumptions and Conditions for Use of the Paired Samples t test

1. The independent variable is dichotomous, and its levels (or groups) are paired, or matched, in some way (e.g., husband-wife, pre-post, etc.).
2. The dependent variable is normally distributed in the two conditions.

The first assumption depends on the design; the second can be assessed by examining the skewness of the two variables.

7.4. Do students' fathers or mothers have more education? Are their education levels correlated?

We will determine if the fathers of these students have more education than their mothers. Remember that the fathers and mothers are paired; that is, each child has a pair of parents whose educations are given in the dataset. (Note that you can do more than one paired *t* test at a time, so we could have compared the *visualization test* and *visualization 2* scores in the same run as we compared *father's* and *mother's education*.)

- Select **Analyze → Compare means and proportions → Paired-Samples T Test**.
- Move <u>both</u> of the variables, *father's education* and *mother's education*, to the **Paired Variables** box (see Figure 7.5).
- Click on **OK**. Compare your syntax and output to Output 7.4.

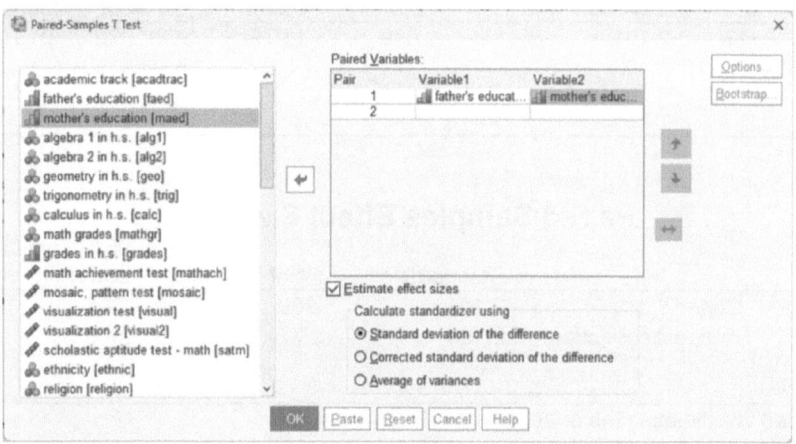

Figure 7.5 Paired-samples t test.

Output 7.4: Paired Samples t Tests

```
T-TEST PAIRS=faed WITH maed (PAIRED)
  /ES DISPLAY(TRUE) STANDARDIZER(SD)
  /CRITERIA=CI(.9500)
  MISSING=ANALYSIS.
```

T-Test

> The circled means are compared.

Paired Samples Statistics

		Mean	N	Std. Deviation	Std. Error Mean
Pair 1	father's education	4.73	73	2.830	.331
	mother's education	4.14	73	2.263	.265

> This is information about the correlation of *mother's education* with *father's education*, not the result of the paired *t*.

Paired Samples Correlations

		N	Correlation	Significance One-Sided p	Two-Sided p
Pair 1	father's education & mother's education	73	.681	<.001	<.001

> The *t* test and its statistical significance.

Paired Samples Test

		Paired Differences							Significance	
					95% Confidence Interval of the Difference					
		Mean	Std. Deviation	Std. Error Mean	Lower	Upper	t	df	One-Sided p	Two-Sided p
Pair 1	father's education - mother's education	.589	2.101	.246	.099	1.079	2.396	72	.010	.019

Paired Samples Effect Sizes

			Standardizer[a]	Point Estimate	95% Confidence Interval Lower	Upper
Pair 1	father's education - mother's education	Cohen's d	2.101	.280	.046	.513
		Hedges' correction	2.123	.277	.045	.508

a. The denominator used in estimating the effect sizes.
Cohen's d uses the sample standard deviation of the mean difference.
Hedges' correction uses the sample standard deviation of the mean difference, plus a correction factor.

Interpretation of Output 7.4

The first table shows the descriptive statistics used to compare *mother's* and *father's education* levels. The second table, **Paired Samples Correlations**, provides correlations between the two paired scores. The correlation ($r = .68$) between *mother's* and *father's education* indicates that highly educated men tend to marry highly educated women and vice versa. It doesn't tell you whether men or women have more education. That is what t in the third table tells you.

The third table shows the **Paired Samples Test**. The **Sig.** for the comparison of the average education level of the students' mothers and fathers was $p = .019$. Thus, the difference in educational level is statistically significant, and we can tell from the means in the first table that fathers have more education; however, as you can see in the final table, the effect size is small ($d = .28$) and is computed by dividing the mean of the paired differences (.59) by the standard deviation (2.1) of the paired differences. Also, we can tell from the confidence interval that the difference in the means could be as small as .10 or as large as 1.08 points on the 2 to 10 scale.

It is important that you understand that the correlation in the second table provides you with different information than the paired t. If not, read this interpretation again.

You should report either "paired" or "correlated."

Always report all this information for statistically significant results.

How to Write About Output 7.4

Results

A paired (or correlated) samples t test indicated that the students' fathers had on average significantly more education, $M = 4.73$, $SD = 2.83$ than their mothers, $M = 4.14$, $SD = 2.26$, $t(72) = 2.40$, $p = .019$, $d = .28$. The difference, although statistically significant, is small using Cohen's (1988) guidelines.

Problem 7.5: Nonparametric Wilcoxon Test for Two Related Samples

Let's assume that education levels and visualization test scores are not normally distributed and/or other assumptions of the paired t test are violated. In fact, *mother's education* was quite skewed (see Chapter 2, Output 2.5a). Let's run the **Wilcoxon signed-ranks** nonparametric test to see if fathers have significantly higher educational levels than the mothers and to see if the *visualization test* is significantly different from the *visualization 2*. The assumptions of the Wilcoxon tests are similar to those for the Mann-Whitney test (see Problem 7.3).

7.5. (a) Are *mother's* and *father's education* levels different?

(b) Are the *visualization test* and *visualization 2* scores different?

- To answer these questions, select **Analyze → Nonparametric Tests → Related Samples**.
- Click on the **Fields** tab.
- Highlight all of the variables in the **Test Fields** box <u>except</u> *father's education* and *mother's education* (if those are in the box) and move them out of the **Test Fields:** box. If needed, move *father's education* and *mother's education* into the box. Only two variables, indicating the same variable for each member of a matched pair, can be included in the box (Figure 7.6).

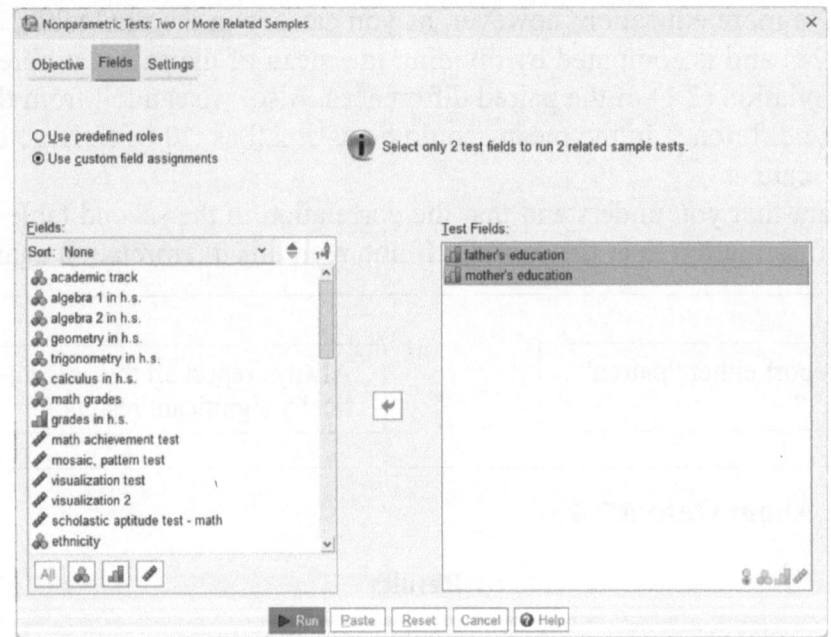

Figure 7.6 Two related-samples tests.

- Click on **Settings**.
- Click on **Wilcoxon matched pair signed rank (2 samples)**.
- Click on **Run**.

Compare your syntax and output to Output 7.5.

Output 7.5: Wilcoxon Nonparametric Test

```
*Nonparametric Tests: Related Samples.
NPTESTS
  /RELATED TEST(faed maed) WILCOXON
  /MISSING SCOPE=ANALYSIS USERMISSING=EXCLUDE
  /CRITERIA ALPHA=0.05 CILEVEL=95.
```

Related-Samples Wilcoxon Signed Rank Test
Summary

Total N	73
Test Statistic	387.500
Standard Error	96.144
Standardized Test Statistic	-2.085
Asymptotic Sig.(2-sided test)	.037

> The test is significant and is not very different in level of significance from the parametric test.

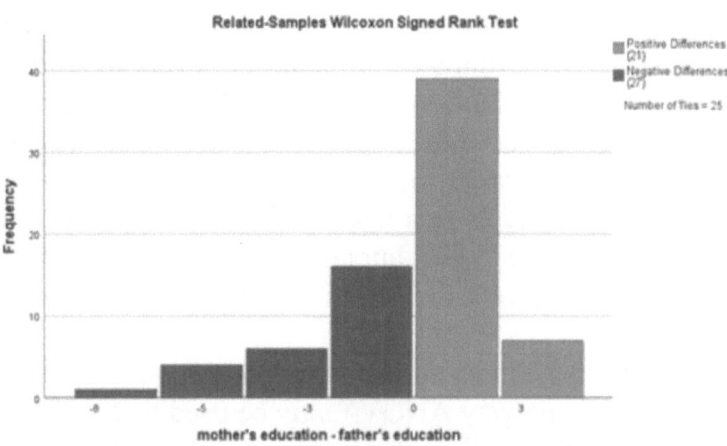

Illustration 7.2 Related-Samples Wilcoxon Signed Rank Test

Interpretation of Output 7.5

Output 7.5 shows the nonparametric (Wilcoxon) analyses, which are similar to the paired *t* tests.

The first table shows the statistical significance level for the two tests. Note that the *p* or Sig. value is similar to that for the paired *t* test comparing father's and mother's education. Effect size measures are not provided on the output, but again we can compute an *r* from the *z* (**Standardized Test Statistic**) scores and *N*s (Total) that are shown in Output 7.5 using the same formula as for Problem 7.3 $(r = \frac{z}{\sqrt{N}})$. For Output 7.5, $r = -.24$ (i.e., $-2.075/8.54$) for the comparison of *mother's* and *father's education*, which is a small effect size. In the figure, you can see the number of students whose mothers had less education than their fathers (27). Note that there were lots of ties (25) and almost as many women (21) that have more education than their husbands. However, overall, the fathers had more education, as indicated by the statistically significant *z* $(p = .037)$.

> It is important to include this information when reporting about ranks.

How to Write About Output 7.5

Results

Wilcoxon signed ranks tests were used to compare the education of each student's mother and father. Of 73 students, 27 fathers had more education, 21 mothers had more education, and there were 25 ties. This difference indicating more education for fathers is statistically significant, $z = -2.09$, $p = .037$, $r = -.24$, a small to medium effect size according to Cohen (1988).

Problem 7.6: One-Way (or Single Factor) ANOVA

In this problem, you will examine a statistical technique for comparing the scores on one dependent variable of more than two groups comprising the levels of a single independent variable. The appropriate statistic, called **One-Way ANOVA**, compares the *means* of the samples or groups in order to make inferences about the population means. One-way ANOVA also is called single factor analysis of variance because there is only one independent variable or factor. The independent variable has nominal levels or a few ordered levels. The overall ANOVA test does not take into account the order of the levels, but additional tests (contrasts) can be done that do consider the order of the levels. More information regarding contrasts can be found in Barrett et al. (2025).

Earlier in this chapter, we used the independent samples *t* test to compare two groups (fast track and regular track students). The one-way ANOVA *may* be used to compare two groups, but it is preferable to use the *t* test to examine differences between two groups using SPSS because: (1) You can easily correct for violations of the equality of variances assumption using the *t* test output but not the ANOVA output and (2) The *t* test output can provide the effect size measure that is best to report for a contrast between two groups, Cohen's *d*. However, ANOVA is necessary if you want to compare three or more groups (e.g., three levels of *father's education*) in a single analysis. Review Table 7.1 to see how these statistics fit into the overall selection of an appropriate statistic.

Assumptions of between-groups/independent samples ANOVA

1. Observations are independent. The value of one observation is not related to any other observation. In other words, one person's score should not provide any clue as to how any of the other people would score. Each person is in only one group and has only one score on each measure; there are no repeated or within-subjects measures.
2. Variances on the dependent variable are equal across groups.
3. The dependent variable is normally distributed for each group.

Because ANOVA is robust, it can be used when variances are only approximately equal if the number of subjects in each group is approximately equal. ANOVA also is robust if the dependent variable data are approximately normally distributed. Thus, if assumption #2, or, even more so, #3 is not fully met, you may still be able to use ANOVA. There are also several choices of post hoc tests to use depending on whether the assumption of equal variances has been violated. **Dunnett's C** and **Games-Howell** are appropriate post hoc tests if the assumption of equal variances is violated.

7.6. Are there statistically significant differences among the three *father's education revised* groups on *grades in h.s.*, *visualization test* scores, and *math achievement*?

We will use the **One-Way ANOVA** procedure because we have one independent variable with three levels. We can do several one-way ANOVAs at a time so we will do three ANOVAs in this problem, one for each of the three dependent variables. Note that you could do MANOVA instead of three ANOVAs, especially if the dependent variables are correlated and conceptually related, but that is beyond the scope of this book. See our companion book Barrett et al. (2025).

To do the three one-way ANOVAs, use the following commands:

- **Analyze → Compare Means and Proportions → One-Way ANOVA...**
- Move *grades in h.s., math achievement,* and *visualization test into* the **Dependent List:** box in Figure 7.7.

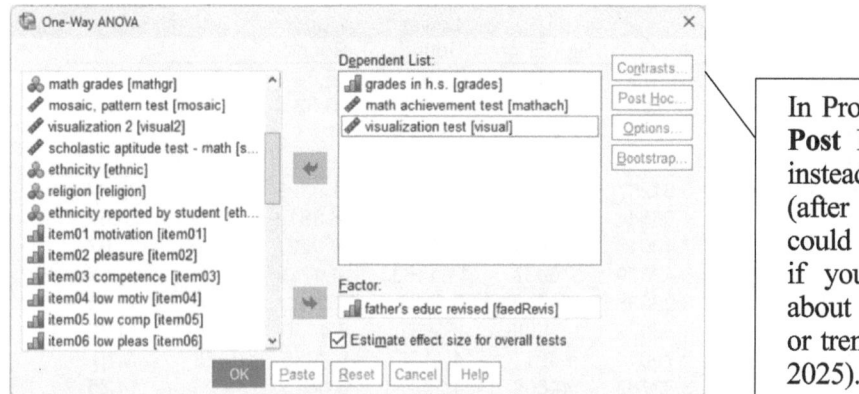

Figure 7.7 One-way ANOVA.

- Click on *father's educ revised* and move it to the **Factor** (independent variable) box.
- Make sure that the **Estimate effect size for overall tests**
- Click on **Options** to get Figure 7.8.
- Under **Statistics**, choose **Descriptive** and **Homogeneity of variance test**.
- Under **Missing Values**, choose **Exclude cases analysis by analysis**.
- Click on **Continue** then **OK**. Compare your output to Output 7.6.

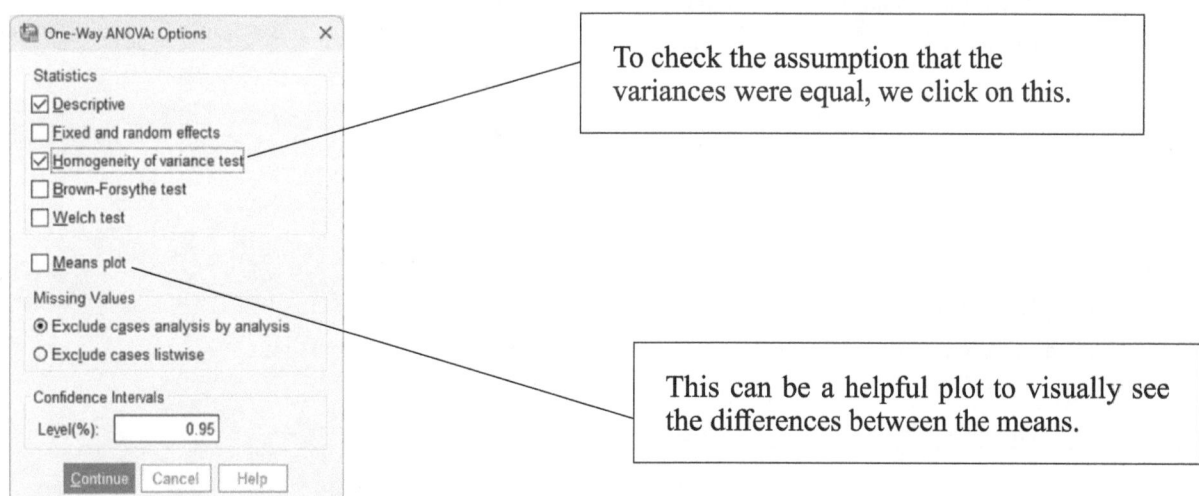

Figure 7.8 One-way ANOVA: Options.

Output 7.6: One-Way ANOVA

```
ONEWAY grades mathach visual BY faedRevis
  /ES=OVERALL
  /STATISTICS DESCRIPTIVES HOMOGENEITY
  /MISSING ANALYSIS
  /CRITERIA=CILEVEL(0.95).
```

Oneway

Means to be compared are circled.

Descriptives

		N	Mean	Std. Deviation	Std. Error	95% Confidence Interval for Mean Lower Bound	Upper Bound	Minimum	Maximum
grades in h.s.	HS grad or less	38	5.34	1.475	.239	4.86	5.83	3	8
	Some College	16	5.56	1.788	.447	4.61	6.52	2	8
	BS or More	19	6.53	1.219	.280	5.94	7.11	4	8
	Total	73	5.70	1.552	.182	5.34	6.06	2	8
math achievement test	HS grad or less	38	10.0877	5.61297	.91054	8.2428	11.9326	1.00	22.67
	Some College	16	14.3958	4.66544	1.16636	11.9098	16.8819	5.00	23.67
	BS or More	19	16.3509	7.40918	1.69978	12.7798	19.9221	1.00	23.67
	Total	73	12.6621	6.49659	.76037	11.1463	14.1779	1.00	23.67
visualization test	HS grad or less	38	4.6711	3.96058	.64249	3.3692	5.9729	-.25	14.75
	Some College	16	6.0156	4.56022	1.14005	3.5857	8.4456	-.25	14.75
	BS or More	19	5.4605	2.79044	.64017	4.1156	6.8055	-.25	9.75
	Total	73	5.1712	3.82787	.44802	4.2781	6.0643	-.25	14.75

Test of Homogeneity of Variances

		Levene Statistic	df1	df2	Sig.
grades in h.s.	Based on Mean	1.546	2	70	.220
	Based on Median	.946	2	70	.393
	Based on Median and with adjusted df	.946	2	67.114	.393
	Based on trimmed mean	1.486	2	70	.233
visualization test	Based on Mean	1.926	2	70	.153
	Based on Median	1.920	2	70	.154
	Based on Median and with adjusted df	1.920	2	64.988	.155
	Based on trimmed mean	1.940	2	70	.151
math achievement test	Based on Mean	3.157	2	70	.049
	Based on Median	1.986	2	70	.145
	Based on Median and with adjusted df	1.986	2	55.820	.147
	Based on trimmed mean	3.046	2	70	.054

Note: This table tests an assumption of ANOVA, not the main hypothesis. The Levene's test is significant for *math achievement* so the variances of the three groups are significantly different, indicating that the assumption is violated. Note that if one based the test on the median, the difference is not significant, suggesting that using a nonparametric test, such as Kruskal-Wallis, would address the problem.

ANOVA

		Sum of Squares	df	Mean Square	F	Sig.
grades in h.s.	Between Groups	18.143	2	9.071	4.091	.021
	Within Groups	155.227	70	2.218		
	Total	173.370	72			
visualization test	Between Groups	22.505	2	11.252	.763	.470
	Within Groups	1032.480	70	14.750		
	Total	1054.985	72			
math achievement test	Between Groups	558.481	2	279.240	7.881	.001
	Within Groups	2480.324	70	35.433		
	Total	3038.804	72			

These are the degrees of freedom: 2, 70.

The between-groups differences for *grades in high school* and *math achievement* are statistically significant ($p < .05$) whereas those for *visualization* are not significantly different.

ANOVA Effect Sizes[a,b]

		Point Estimate	95% Confidence Interval Lower	Upper
grades in h.s.	Eta-squared	.105	.001	.235
	Epsilon-squared	.079	-.027	.213
	Omega-squared Fixed-effect	.078	-.027	.211
	Omega-squared Random-effect	.041	-.013	.118
math achievement test	Eta-squared	.184	.038	.325
	Epsilon-squared	.160	.010	.305
	Omega-squared Fixed-effect	.159	.010	.302
	Omega-squared Random-effect	.086	.005	.178
visualization test	Eta-squared	.021	.000	.105
	Epsilon-squared	-.007	-.029	.080
	Omega-squared Fixed-effect	-.007	-.028	.079
	Omega-squared Random-effect	-.003	-.014	.041

a. Eta-squared and Epsilon-squared are estimated based on the fixed-effect model.
b. Negative but less biased estimates are retained, not rounded to zero.

Interpretation of Output 7.6

The first table, **Descriptives**, provides familiar descriptive statistics for the three father's education groups on each of the three dependent variables (*grades in h.s.*, *visualization test*, and *math achievement*) that we requested for these analyses. Remember that, although these three dependent variables appear together in each of the tables, we have really computed three separate one-way ANOVAs.

The second table (**Test of Homogeneity of Variances**) provides the Levene's test to check the assumption that the variances of the three *father's education* groups are equal for each of the dependent variables. For ANOVA, we are interested in the test based on means. Notice that for *grades in h.s.* ($p = .220$) and *visualization test* ($p = .153$) the Levene's tests are *not* significant. Thus, the assumption is *not* violated. However, for *math achievement, p =* .049; therefore, the Levene's test is significant, and thus the assumption of equal variances is violated. In this latter case, we could use the similar nonparametric test (Kruskal-Wallis). Or if the overall F is significant (as you can see it was in the ANOVA table), you could use a post hoc test designed for situations in which the variances are unequal. We will do the latter in Problem 7.7b and the former (i.e., the Kruskal-Wallis test) in Problem 7.8 for *math achievement*.

The **ANOVA** table in Output 7.6 is the key table because it shows whether the overall Fs for these three ANOVAs were statistically significant. Note that the three *father's education* groups differ on *grades in h.s.* and *math achievement* but not *visualization test*. When reporting these findings one should write, for example, $F(2, 70) = 4.09, p = .021$, for *grades in h.s.* The 2, 70 (circled for *grades in h.s.* in the ANOVA table) are the degrees of freedom (*df*) for the between-groups "effect" and the within-groups "error," respectively. *F tables* also usually include the mean squares, which indicate the amount of variance (sums of squares) for that "effect" divided by the degrees of freedom for that "effect." You also should report the means (and SDs) so that one can see which groups were high and low.

The final table provides estimates of overall effect size. Eta squared is widely used and easy to interpret, in that it indicates the proportion of variance in the dependent variable that is due to differences between the independent variable groups. It is also the best measure to use if you want to compare different studies. However, it has statistical limitations in relation to how it estimates the population effect size that are beyond the scope of this book. If you want a measure that does not have this problem, especially if you have a small sample, you can report Omega Squared. However, it is less intuitive in its interpretation. Regardless, remember that this just tells you that there is this amount of difference somewhere among your groups; if you have three or more groups, you will not know which specific pairs of means are significantly different unless you do a priori (beforehand) contrasts (see Figure 7.7) or post hoc tests, as shown next in Problem 7.7. We provide an example of appropriate APA-format tables and how to write about these ANOVAs after Problem 7.7.

Problem 7.7: Post Hoc Multiple Comparison Tests

Now we will introduce the concept of **post hoc multiple comparisons**, sometimes called **follow-up tests**. When you compare three or more group means, you know that there will be a statistically significant difference somewhere if the **ANOVA** F (sometimes called the **overall** F or **omnibus** F) is significant.

However, we would usually like to know which specific means are different from which other ones. In order to know this, you can use *one* of several post hoc tests that are built into the one-way ANOVA program. The **LSD** post hoc test is quite liberal (most would say it is too lenient), and the **Scheffe** test is quite conservative, so many statisticians recommend a more middle of the road test, such as the **Tukey HSD** (honestly significant differences) test if the Levene's test was <u>not</u> statistically significant, or the **Games-Howell** test if the Levene's test was significant. Ordinarily, you <u>only do post hoc tests only if the overall *F* is statistically significant</u>. For this reason, we have separated Problems 7.6 and 7.7, which could have been done in one step. Figure 7.9 shows the steps one should use in deciding whether to use post hoc multiple comparison tests.

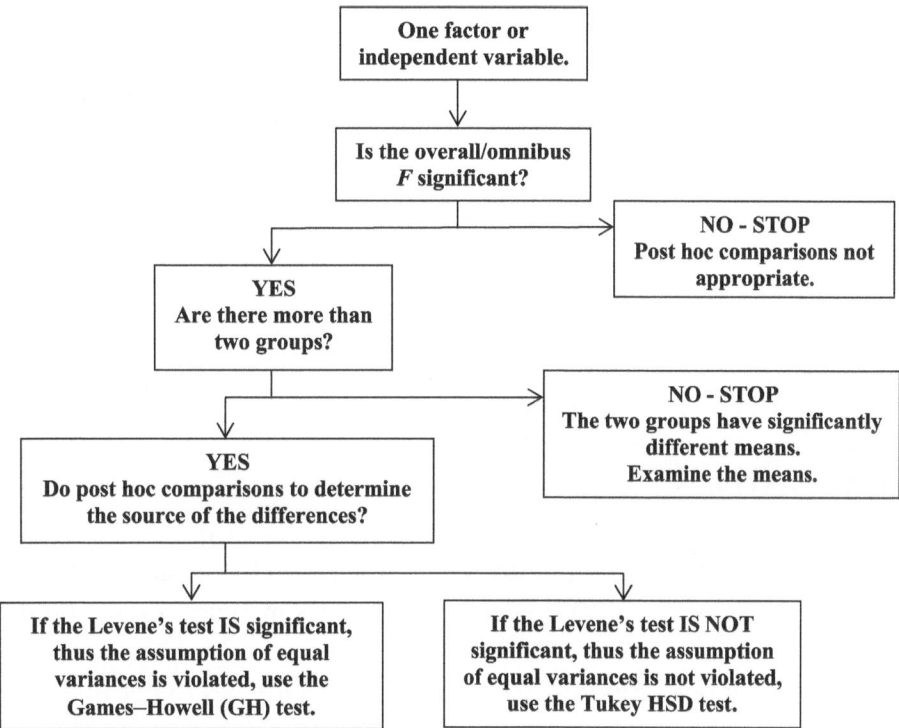

Figure 7.9 Schematic representation of when to use post hoc multiple comparisons with a one-way ANOVA.

7.7. When the overall *F* is statistically significant, which pairs of means are significantly different?

After you have examined Output 7.6 to see if the overall *F* (ANOVA) for each variable was statistically significant, you will do appropriate post hoc multiple comparisons for these variables. We will use the **Tukey HSD** if variances can be assumed to be equal (i.e., the Levene's test is *not* statistically significant) and the **Games-Howell** if the assumption of equal variances cannot be justified (i.e., the Levene's test is significant).

First we will do the Tukey HSD for *grades in h.s.* Open the **One-Way ANOVA** dialog box *again* by doing the following:

- Select **Analyze → Compare Means and Proportions → One-Way ANOVA . . .** to see Figure 7.7 again.
- Move *visualization test* out of the **Dependent List** box by highlighting it and clicking on the arrow pointing left because the overall *F* for *visualization test* was not significant. (See interpretation of Output 7.6.)
- Also move *math achievement* to the left (out of the **Dependent List** box) because the Levene's test for it <u>was</u> statistically significant. (We will use it later.)

- Keep *grades* in the **Dependent List** box because it had a significant ANOVA, and the Levene's test was not statistically significant.
- Ensure that *father's educ revised* is in the **Factor** box.
- Your window should look like Figure 7.10.

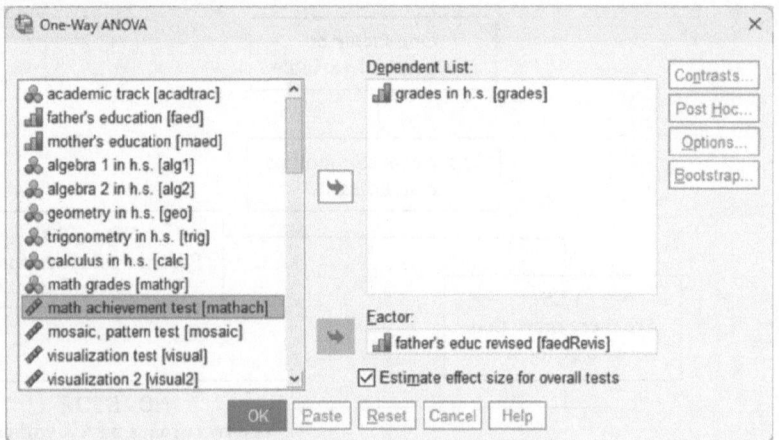

Figure 7.10 One-Way ANOVA.

- Next, click on **Options . . .** and remove the check for **Descriptive** and **Homogeneity of variance test** (in Figure 7.8) because we do not need to do them again; they would be the same.
- Click on **Continue**.
- Then, in the main dialog box (Figure 7.7), click on **Post Hoc . . .** to get Figure 7.11.
- Check **Tukey** because, for *grades in h.s.*, the Levene's test was not significant, so we can assume that the variances are approximately equal.

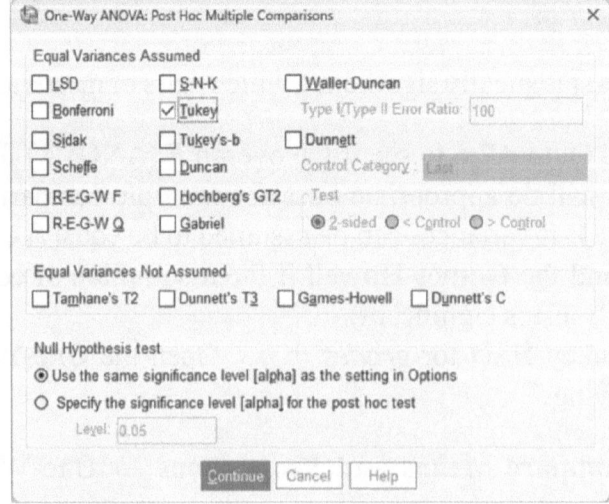

Figure 7.11 One-way ANOVA: Post hoc multiple comparisons.

- Click on **Continue** and then **OK** to run this post hoc test.

Compare your output to Output 7.7a.

Output 7.7a: Tukey HSD Post Hoc Tests

```
ONEWAY grades BY faedRevis
  /ES=OVERALL
  /MISSING ANALYSIS
  /CRITERIA=CILEVEL(0.95)
  /POSTHOC=TUKEY ALPHA(0.05).
```

Oneway

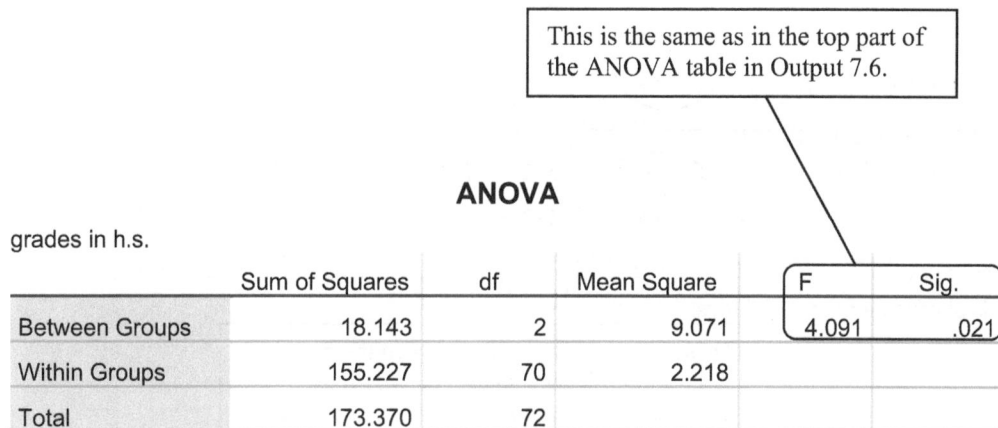

This is the same as in the top part of the ANOVA table in Output 7.6.

ANOVA

grades in h.s.

	Sum of Squares	df	Mean Square	F	Sig.
Between Groups	18.143	2	9.071	4.091	.021
Within Groups	155.227	70	2.218		
Total	173.370	72			

ANOVA Effect Sizes[a,b]

		Point Estimate	95% Confidence Interval	
			Lower	Upper
grades in h.s.	Eta-squared	.105	.001	.235
	Epsilon-squared	.079	-.027	.213
	Omega-squared Fixed-effect	.078	-.027	.211
	Omega-squared Random-effect	.041	-.013	.118

a. Eta-squared and Epsilon-squared are estimated based on the fixed-effect model.
b. Negative but less biased estimates are retained, not rounded to zero.

Post Hoc Tests

> The Tukey HSD is a common post hoc test to use when variances are equal. This output is most appropriate when the group sizes (*n*) are similar. Here they are quite different. Therefore it is better to use the Homogeneous Subset table.

Multiple Comparisons

Dependent Variable:　grades in h.s.

Tukey HSD

(I) father's educ revised	(J) father's educ revised	Mean Difference (I-J)	Std. Error	Sig.	95% Confidence Interval Lower Bound	95% Confidence Interval Upper Bound
HS grad or less	Some College	-.220	.444	.873	-1.28	.84
	BS or More	-1.184*	.418	.017	-2.19	-.18
Some College	HS grad or less	.220	.444	.873	-.84	1.28
	BS or More	-.964	.505	.144	-2.17	.25
BS or More	HS grad or less	1.184*	.418	.017	.18	2.19
	Some College	.964	.505	.144	-.25	2.17

*. The mean difference is significant at the 0.05 level.

> These are the differences between the means and the significance levels you would use if the group sizes were similar. Ignore the duplicates. (We have put lines through them.)

Homogeneous Subsets

Tukey HSD[a,b]

father's educ revised	N	Subset for alpha = 0.05 — 1	Subset for alpha = 0.05 — 2
HS grad or less	38	5.34	
Some College	16	5.56	5.56
BS or More	19		6.53
Sig.		.880	.096

> The harmonic mean is the mean of the groups' sample sizes

Means for groups in homogeneous subsets are displayed.
a. Uses Harmonic Mean Sample Size = 21.209.
b. The group sizes are unequal. The harmonic mean of the group sizes is used. Type I error levels are not guaranteed.

> This way of computing and displaying the post hoc tests is more appropriate when group sizes are quite different. Groups listed in the same subset are not significantly different. Thus, the grades of students whose fathers were *HS grads or less* are not different from those whose fathers had *some college*. Likewise, those with *some college* are not different from those with a *BS or more*, but *HS grads or less* are different from those with a *BS or more*. In this case, the conclusion is the same as from the **Post Hoc Tests** table.

Interpretation of Output 7.7a

The first table in both Outputs 7.7a and 7.7b repeats appropriate parts of the **ANOVA** table from Output 7.6. The second table in Output 7.7a shows the **Tukey HSD** test for *grades in h.s.* that you would use if the three group sizes (n = 38, 16, 19) had been similar. For *grades in h.s.*, this Tukey table indicates that there is only a small not statistically significant mean difference (.22) between the mean grades of students whose fathers were *high school grads or less* (M = 5.34 from Output 7.6) and those fathers who had *some college* (M = 5.56). The **Homogeneous Subsets** table shows an <u>adjusted Tukey</u> that is appropriate when group sizes are not similar, as in this case. It bases this on the **harmonic mean**, which is the mean of the group sample sizes. Note that there is <u>not</u> a statistically significant difference (p = .880) between the grades of students whose fathers were *high school grads or less* (low education) and those with *some college* (medium education) because <u>their means are both shown in</u> **Subset 1**. In **Subset 2**, the medium and high education group means are shown, indicating that they are not significantly different (p = .096). By examining the two subset boxes, we can see that the low education group (M = 5.34) is different from the high education group (M = 6.53) because these two means do not appear in the same subset.

After you do the Tukey test, let's go back and do **Games-Howell** for math achievement because equal variances cannot be assumed. Follow these steps:

- Select **Analyze → Compare Means and Proportions → One-Way ANOVA . . .**
- Move *grades in h.s.* out of the **Dependent List** box by highlighting it and clicking on the arrow pointing left.
- Move *math achievement* into the **Dependent List** box.
- Ensure that *father's educ revised* is still in the **Factor** box.
- In the main dialog box (Figure 7.7), click on **Post Hoc . . .** to get Figure 7.10.
- Check **Games-Howell** because equal variances cannot be assumed for *math achievement*.
- <u>Remove</u> the check mark from Tukey.
- Click on **Continue** and then **OK** to run this post hoc test.
- Compare your syntax and output to Output 7.7b.

Output 7.7b: Games-Howell Post Hoc Test

```
ONEWAY mathach BY faedRevis
  /ES=OVERALL
  /MISSING ANALYSIS
  /CRITERIA=CILEVEL(0.95)
  /POSTHOC=GH ALPHA(0.05).
```

Oneway

ANOVA

math achievement test

	Sum of Squares	df	Mean Square	F	Sig.
Between Groups	558.481	2	279.240	7.881	.001
Within Groups	2480.324	70	35.433		
Total	3038.804	72			

ANOVA Effect Sizes[a]

		Point Estimate	95% Confidence Interval	
			Lower	Upper
math achievement test	Eta-squared	.184	.038	.325
	Epsilon-squared	.160	.010	.305
	Omega-squared Fixed-effect	.159	.010	.302
	Omega-squared Random-effect	.086	.005	.178

a. Eta-squared and Epsilon-squared are estimated based on the fixed-effect model.

Post Hoc Tests

Multiple Comparisons

We used Games–Howell because the Levene's test indicated that the variances are unequal.

Dependent Variable: math achievement test

Games-Howell

(I) father's educ revised	(J) father's educ revised	Mean Difference (I-J)	Std. Error	Sig.	95% Confidence Interval	
					Lower Bound	Upper Bound
HS grad or less	Some College	-4.30810*	1.47969	.017	-7.9351	-.6811
	BS or More	-6.26324*	1.92830	.008	-11.0284	-1.4980
Some College	HS grad or less	4.30810*	1.47969	.017	.6811	7.9351
	BS or More	-1.95513	2.06147	.614	-7.0308	3.1205
BS or More	HS grad or less	6.26324*	1.92830	.008	1.4980	11.0284
	Some College	1.95513	2.06147	.614	-3.1205	7.0308

*. The mean difference is significant at the 0.05 level.

Interpretation of Output 7.7b

The first table shows the same ANOVA results we computed in 7.7a. The second table presents the **Games-Howell** test, used for variables that have unequal variances. Note that each comparison is presented twice (so we have marked through three of the lines where they are redundant—do not look at these). Pay attention to the three lines that have circles.

The **Mean Difference** on math achievement between students whose fathers were *high school grads or less* and those with fathers who had *some college* was –4.31. The **Sig.**

(p = .017) indicates that this is a statistically significant difference. We can also tell that this difference is statistically significant because the confidence interval's lower and upper bounds both have the same sign, a minus, so zero (no difference) is not included in the confidence interval. Similarly, students whose fathers had a *B.S. degree* were significantly different on *math achievement* from those whose fathers had a *high school degree or less* (p = .008).

It is important to always include the direction of the results (i.e., which group had the higher mean).

Example of How to Write About Outputs 7.6 and 7.7

Results

A statistically significant difference was found among the three levels of father's education for grades in high school, $F(2, 70) = 4.09$, $p = .021$, and math achievement, $F(2, 70) = 7.88$, $p = .001$. Table 7.4 shows that the mean grade in high school is 5.34 for students whose fathers had low education, 5.56 for students whose fathers attended some college (medium), and 6.53 for students whose fathers received a BS or more (high). Post hoc Tukey HSD tests were used to follow significant overall ANOVAs if there were homogeneous variances and Games-Howell if there were heterogeneous variances. Tukey HSD tests indicated that the low education group and high education group differed significantly in their grades with a large effect size ($p < .05$, $d = .85$). Likewise, there were also statistically significant mean differences on math achievement between the low education and both the medium education group ($p < .017$, $d = .80$) and the high education group ($p = .008$, $d = 1.0$) using the Games-Howell post hoc test.

Include which post hoc test you conducted.

Recall that you need to include the means and standard deviations in order for readers to compute effect sizes.

Table 7.4
Means and Standard Deviations Comparing Three Father's Education Groups

Father's education	n	Grades in H.S. M	SD	Math achievement M	SD	Visualization Father's education M	SD
HS grad or less (low)	38	5.34	1.48	10.09	5.61	4.67	3.96
Some college (medium)	16	5.56	1.79	14.40	4.67	6.02	4.56
BS or more (high)	19	6.53	1.22	16.35	7.41	5.46	2.79
Total	73	5.70	1.55	12.66	6.50	5.17	3.83

> A table with this information is typical when reporting ANOVA results.

Table 7.5

One-Way Analysis of Variance Summary Table Comparing Father's Education Groups on Grades in High School, Math Achievement, and Visualization Test

Source	df	SS	MS	F	p
Grades in high school					
Between groups	2	18.14	9.07	4.09	.021
Within groups	70	155.23	2.22		
Total	72	173.37			
Math achievement					
Between groups	2	558.48	279.24	7.88	.001
Within groups	70	2480.32	35.43		
Total	72	3038.80			
Visualization test					
Between groups	2	22.51	11.25	.76	.470
Within groups	70	1032.48	14.75		
Total	72	1054.99			

Problem 7.8: Nonparametric Kruskal-Wallis Test

What else can you do if the homogeneity of variance assumption is violated (or if your data are ordinal or highly skewed)? One answer is a nonparametric statistic. Let's make comparisons similar to Problem 7.6, assuming that the data are ordinal or the assumption of equality of group variances is violated. Remember that the variances for the three father's education groups were statistically significantly different on *math achievement*, and the *competence scale* was not normally distributed (see Chapter 2). The assumptions of the Kruskal-Wallis test are the same as for the Mann-Whitney test (earlier in this chapter).

7.8. Are there statistically significant differences among the three father's education groups on *math achievement* and the *competence scale*?

Follow these commands:

- **Analyze → Nonparametric Tests → Independent Samples . . .**
- Click on the **Fields** tab
- Move all of the dependent variables except *math achievement* and *competence* out of the **Test Fields** box: (see Figure 7.12).
- Move the independent variable *father's educ revised* to the **Groups** box.
- Then click on **RUN**. Do your results look like Output 7.8?

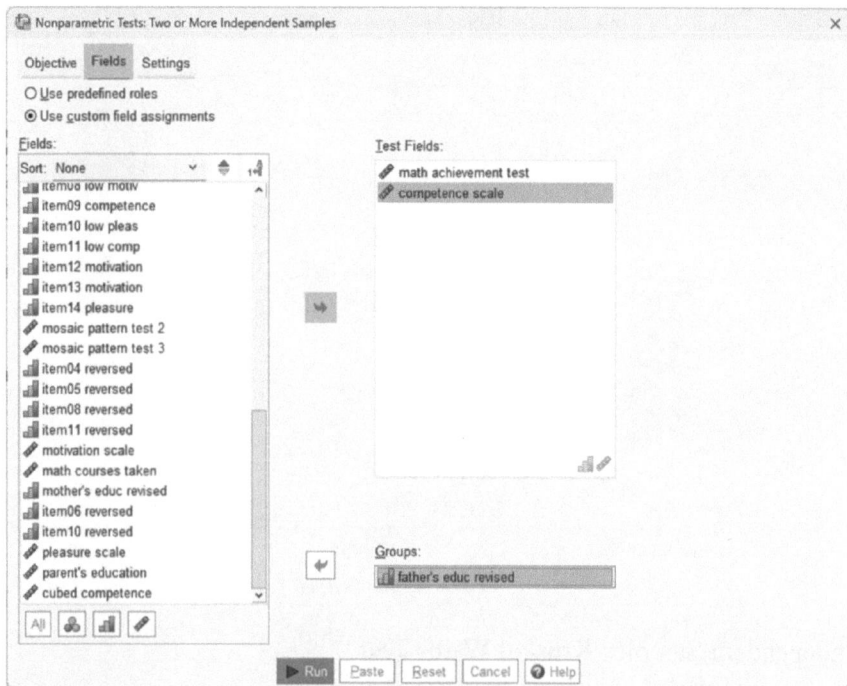

Figure 7.12 Tests for several independent samples.

Output 7.8: Kruskal-Wallis Nonparametric Tests

```
*Nonparametric Tests: Independent Samples. NPTESTS
  /INDEPENDENT TEST (mathach competence) GROUP (faedRevis)
  /MISSING SCOPE=ANALYSIS USERMISSING=EXCLUDE
  /CRITERIA ALPHA=0.05 CILEVEL=95.
```

Independent-Samples Kruskal-Wallis Test Math Achievement Test Across Father's Educ Revised

Independent-Samples Kruskal-Wallis Test Summary

Total N	73
Test Statistic	13.384[a]
Degree Of Freedom	2
Asymptotic Sig.(2-sided test)	.001

a. The test statistic is adjusted for ties.

Pairwise Comparisons of father's educ revised

Sample 1-Sample 2	Test Statistic	Std. Error	Std. Test Statistic	Sig.	Adj. Sig.[a]
HS grad or less-Some College	-15.347	6.311	-2.432	.015	.045
HS grad or less-BS or More	-19.987	5.950	-3.359	<.001	.002
Some College-BS or More	-4.640	7.186	-.646	.518	1.000

Each row tests the null hypothesis that the Sample 1 and Sample 2 distributions are the same.
Asymptotic significances (2-sided tests) are displayed. The significance level is .050.
a. Significance values have been adjusted by the Bonferroni correction for multiple tests.

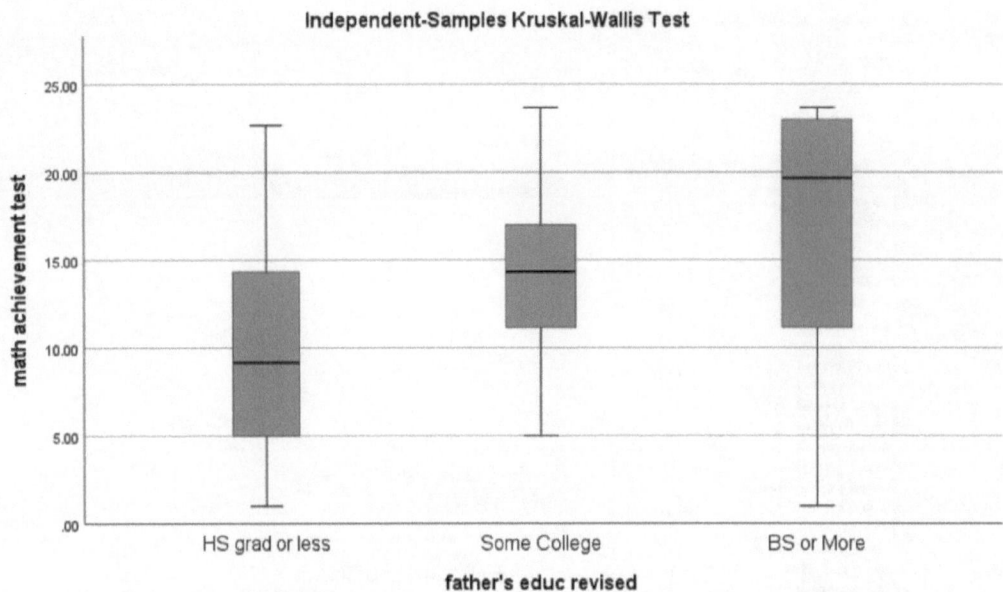

Illustration 7.3 Independent Samples Kruskal Wallis Test

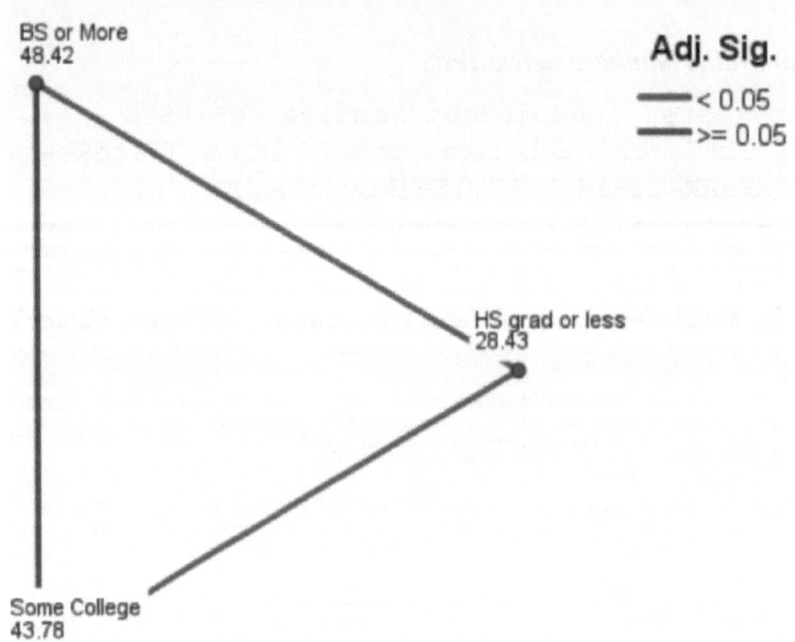

Illustration 7.4 Pairwise Comparisons of Father's Education Revised

competence scale across father's educ revised

Independent-Samples Kruskal-Wallis Test Summary

Total N	71
Test Statistic	.003[a]
Degree Of Freedom	2
Asymptotic Sig.(2-sided test)	.999

a. The test statistic is adjusted for ties.

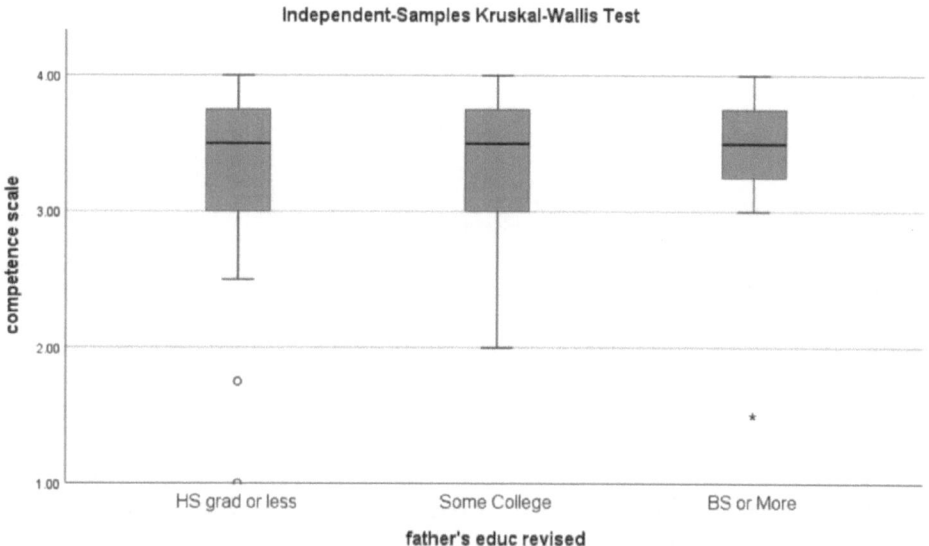

Illustration 7.5 Independent Samples Kruskal-Wallis Test—Competence Scale

Pairwise Comparisons of father's educ revised

Sample 1-Sample 2	Test Statistic	Std. Error	Std. Test Statistic	Sig.	Adj. Sig.[a]
Some College-HS grad or less	.259	6.085	.043	.966	1.000
Some College-BS or More	-.330	6.987	-.047	.962	1.000
HS grad or less-BS or More	-.071	5.844	-.012	.990	1.000

Each row tests the null hypothesis that the Sample 1 and Sample 2 distributions are the same.
Asymptotic significances (2-sided tests) are displayed. The significance level is .050.
a. Significance values have been adjusted by the Bonferroni correction for multiple tests.

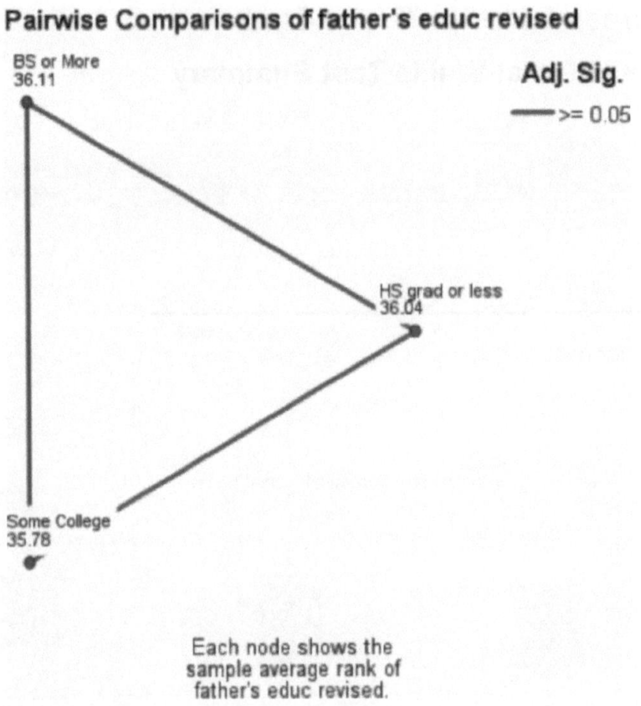

Pairwise Comparisons of father's educ revised

Illustration 7.6 Pairwise Comparisons of Father's Education Revised Competence scale

---

Interpretation of Output 7.8

As in the case of the Mann-Whitney test (Problem 7.3), the **Kruskal-Wallis** (K-W) test compares the mean ranks for the three *father's education* groups.

The **Test Statistics** table shows whether there is an overall difference among the three groups. The Kruskal-Wallis test is listed here as "Test statistic" but should be reported as χ^2. Notice that the *p* (Asymptotic Sig.) value for *math achievement* is .001, which is the same as it was in Output 7.6 using the one-way ANOVA. This is because K-W and ANOVA have similar power to detect a difference. Note also that there is not a statistically significant difference among the father's education groups on the *competence scale* ($p = .999$).

The next thing you see in the output is box charts that show you how the groups compare, not only on their mean ranks/medians (depicted as the horizontal black line in the boxes), but also on their distribution. The next table shows you pairwise post hoc tests built into the K-W test, which have been corrected for multiple comparisons. Thus, you can tell which of the pairs of father's education means are different on *math achievement*. Note you would only do the post hoc M-W tests if the K-W test was statistically significant; thus, you should not report the pairwise comparisons for *competence*. The significance level was adjusted by dividing .05 by 3 (the Bonferroni correction) so that the *p* value to achieve significance would be < .017. Finally, there is a graphic depiction of which pairwise comparisons were significant. This figure also provides the mean rank for each group.

How to Write About Output 7.8

Results

A Kruskal-Wallis nonparametric test was conducted to test for statistically significant differences between father's education groups in math achievement because there were unequal variances and *n*s across groups. The test indicated that the three father's education groups differed on math achievement, χ^2 (2, $N = 73$) = 13.38, $p = .001$. Post hoc pairwise comparisons compared the three pairs of father's education groups on math achievement, using a Bonferroni corrected *p* value of .017 to indicate statistical significance. The mean rank for math achievement of students whose fathers had some college (43.78, $n = 16$) was significantly higher than that of students whose fathers were high school graduates or less (28.43, $n = 38$), $z = -2.43$, $p = .015$, $r = .33$, a medium to large effect size according to Cohen (1988). Also, the mean rank for math achievement of students whose fathers had a bachelor's degree or more (48.42, $n = 19$) was higher than that of students whose fathers were high school graduates or less (28.43, $n \ne 38$), $z = -3.36$, $p = .001$, $r = .46$, a statistically significant difference with a medium to large effect size. There was not a statistically significant difference in math achievement between students whose fathers had some college and those whose fathers had a bachelor's degree or more, $z = -.646$, $p = .52$, $r = .09$. The effect size for this comparison was small.

> Remember to always interpret the effect size.

Problem 7.9: Two-Way (or Factorial) ANOVA

In previous problems, we compared two or more groups based on the levels of only one independent variable or factor using *t* tests and one-way ANOVA. These were called single factor designs. In this problem, we will compare groups based on *two* independent variables. The appropriate statistic for this is called a two-way or factorial ANOVA. This statistic is used when there are two different independent variables, each of which classifies participants with respect to a particular characteristic, with each participant being in one level of each of the independent variables (completely crossed design). For example, on the variables of academic track and math grades, one student could be on the regular track and have high math grades (i.e., mostly A–B). Another student might be on the fast track and have lower math grades (less than A–B). Every student in the HSB dataset who is not missing data on these variables has some combination of one of the two levels of math grades and be on either fast or regular track in high school. In this chapter, we provide an introduction to this complex difference statistic; a more in-depth treatment is provided in Barrett et al. (2025).

7.9. Do *math grades* and *academic track* each seem to have a statistically significant "effect" on *math achievement*, and do the "effects" of *math grades* on *math achievement* depend on whether the person is in the regular or fast track (i.e., on the interaction of *math grades* with *academic track*)?

Follow these commands:

- **Analyze → General Linear Model → Univariate . . .**
- Move *math achievement* to the **Dependent Variable** box.
- Move the first independent variable, *math grades* (<u>not *grades in h.s.*</u>), to the **Fixed Factor(s):** box.
- Then also move the second independent variable, *academic track*, to the **Fixed Factor(s):** box (see Figure 7.13).

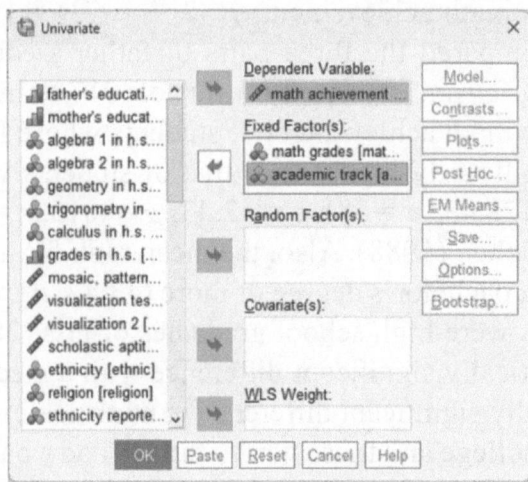

Figure 7.13 GLM Univariate.

Now that we know the variables we will be dealing with, let's determine our options.

- Click on **Plots** and move *mathgr* to the **Horizontal Axis** and *acadtrac* to the **Separate Lines** box.
- Then press **Add**. Your window should <u>now</u> look like Figure 7.14.
- Click on **Continue** to get back to Figure 7.13.

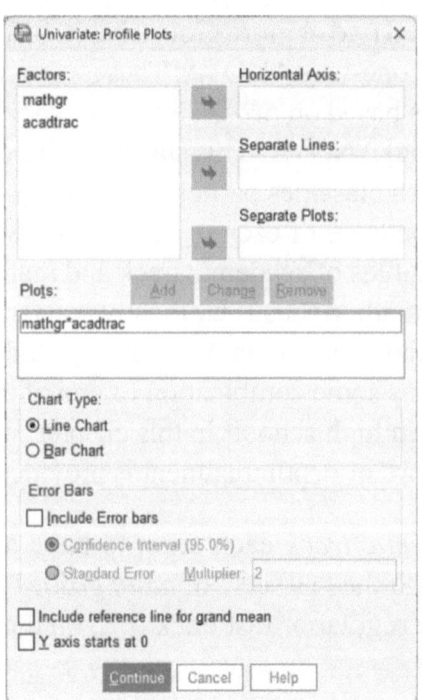

Figure 7.14 Univariate: Profile plots.

- Select **Options** and click **Descriptive statistics** and **Estimates of effect size** (see Figure 7.15).
- Leave the **Significance level** as .05 (see Figure 7.15).
- Click on **Continue**.
- Click on **OK**. Compare your syntax and output to Output 7.9.

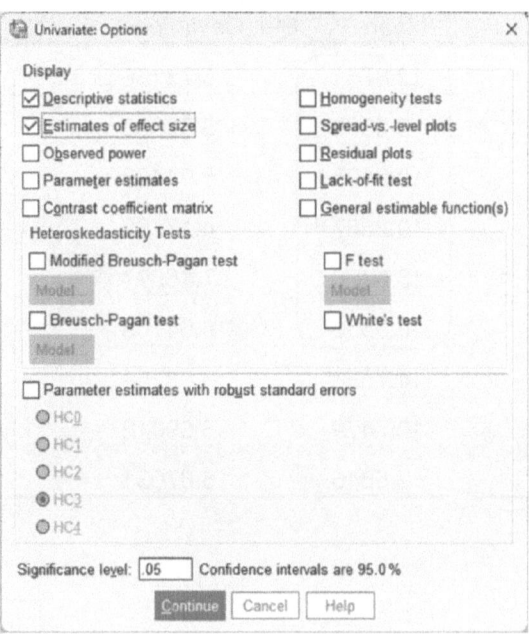

Figure 7.15 Univariate: Options.

Output 7.9: Two-Way ANOVA

```
UNIANOVA mathach BY mathgr acadtrac
  /METHOD=SSTYPE(3)
  /INTERCEPT=INCLUDE
  /PLOT=PROFILE(mathgr*acadtrac) TYPE=LINE ERRORBAR=NO
  MEANREFERENCE=NO YAXIS=AUTO
  /PRINT ETASQ DESCRIPTIVE
  /CRITERIA=ALPHA(.05)
  /DESIGN=mathgr acadtrac mathgr*acadtrac.
```

Univariate Analysis of Variance
Between-Subjects Factors

		Value Label	N
math grades	0	less A-B	44
	1	most A-B	31
academic track	0	fast track	34
	1	regular track	41

Descriptive Statistics

Dependent Variable: math achievement test

math grades	academic track	Mean	Std. Deviation	N
less A-B	fast track	12.8751	5.73136	24
	regular track	8.3333	5.32563	20
	Total	10.8106	5.94438	44
most A-B	fast track	19.2667	4.17182	10
	regular track	13.0476	7.16577	21
	Total	15.0538	6.94168	31
Total	fast track	14.7550	6.03154	34
	regular track	10.7479	6.69612	41
	Total	12.5645	6.67031	75

> The cell means are important for interpreting factorial ANOVAs and describing the results.

> These *F*s and significance levels tell you important information about differences between means and the interaction.

Tests of Between-Subjects Effects
Dependent Variable: math achievement test

Source	Type III Sum of Squares	df	Mean Square	F	Sig.	Partial Eta Squared
Corrected Model	814.481[a]	3	271.494	7.779	<.001	.247
Intercept	11971.773	1	11971.773	343.017	<.001	.829
mathgr	515.463	1	515.463	14.769	<.001	.172
acadtrac	483.929	1	483.929	13.866	<.001	.163
mathgr * acadtrac	11.756	1	11.756	.337	.563	.005
Error	2478.000	71	34.901			
Total	15132.393	75				
Corrected Total	3292.481	74				

a. R Squared = .247 (Adjusted R Squared = .216)

> Percent of variance in *math achievement* predictable from both independent variables and the interaction.

> Eta squared is an index of the effect size for each independent variable and the interaction. Thus, about 17% of the variance in *math achievement* can be predicted from *math grades*.

Profile Plots

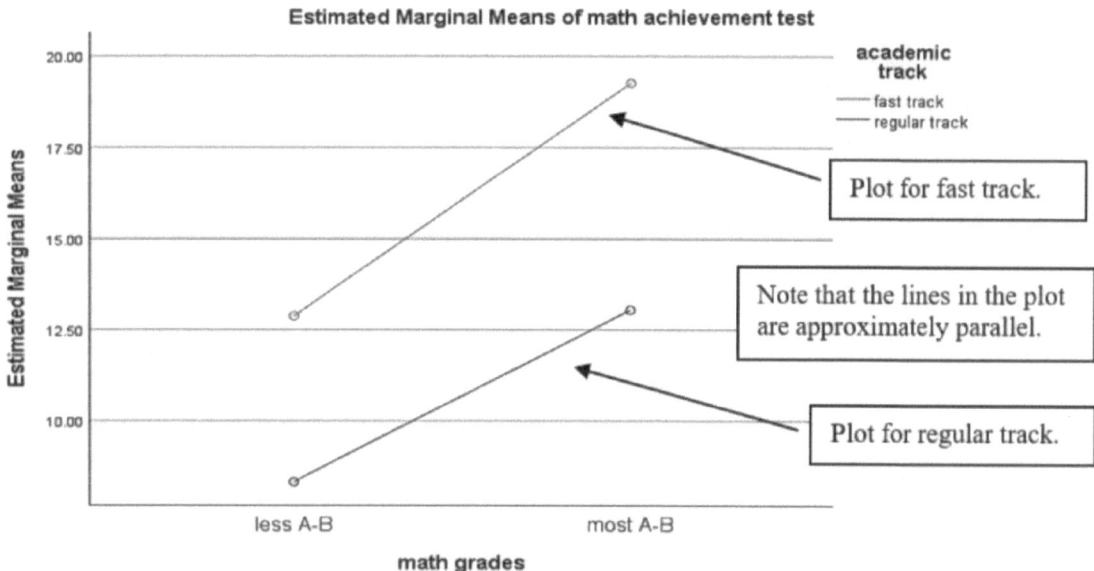

Illustration 7.7 Profile Plots Math Grades

Interpretation of Output 7.9

The GLM Univariate program allows you to print the means for each subgroup (cell) representing the interaction between the two independent variables. It also provides measures of effect size (eta²) and plots the interaction, which is helpful in interpreting it. The first table in Output 7.9 shows that 75 participants (44 with less than A–B math grades and 31 mostly A–B math grades) are included in the analysis because they had data on all of the three variables. The **Descriptive Statistics** table shows the cell and marginal (total) means; both are very important for interpreting the ANOVA table and explaining the results of the test for the interaction.

The ANOVA table, called **Tests of Between-Subjects Effects**, is the key table. Note that the word "effect" in the title of the table can be misleading because this study was not a randomized experiment. Thus, you cannot say in your report that the differences in the dependent variable were *caused* by or were the effect of the independent variable. Usually you will ignore the lines in the table labeled "corrected model" (which just summarizes all "effects" taken together) and intercept (which is needed to fit the best fit regression line to the data) and skip down to the interaction *F*(***mathgr * academic track***), which, in this case, is not statistically significant, *F*(1, 71) = .337, *p* = .563. If the interaction were significant, we would need to be cautious about the interpretation of the main effects because they could be misleading.

If the interaction is statistically significant, you should also analyze the differences between cell means (the simple effects). Barrett et al. (2025) shows how to do this and discusses more about how to interpret statistically significant interactions. The profile plots may be helpful in visualizing the interaction, but you should not discuss statistically nonsignificant differences because the plots may be misleading.

Next we examine the main effects of *math grades* and of *academic track*. Note that both are statistically significant. The significant *F* for *math grades* means that students with fewer As and Bs in math scored lower on *math achievement* than those with high math grades ($M =$ 10.81 vs. 15.05), and this difference is statistically significant ($p < .001$). *Academic track* is also statistically significant ($p < .001$). Because the interaction is not statistically significant, the "effect" of *math grades* on *math achievement* is about the same for both academic tracks. If the interaction were statistically significant, we would say that the "effect" of math grades depended on which academic track you were considering. For example, it might be large for the fast track and small for the regular track.

Note also the callout boxes about the adjusted *R* squared and partial eta squared. Eta, the correlation ratio, is used when the independent variable is *nominal* and the dependent variable (*math achievement* in this problem) is normal (i.e., approximately normally distributed). Eta^2 is an indicator of the proportion of variance that is due to between-groups differences. Because eta and *R*, like *r*, are indexes of association, they can be used as measures of the effect size. See Table 7.2 for Cohen's guidelines for small, medium, and large are somewhat different (for eta, small = .10, medium = .24, and large = .37).

In this example, eta (not squared) for *math grades* is about .41 and thus a medium to large effect. Eta for *academic track* is about .40, also a medium to large effect. The overall adjusted *R* is about .46, a large effect. Notice that the adjusted R^2 is lower than the unadjusted (.22 versus .25). The reason for this is that the adjusted R^2 takes into account (and adjusts for) the fact that not just one variable but three (*math grades*, *academic track*, and the interaction) were used to predict *math achievement*.

An important point to remember is that statistical significance depends heavily on the sample size, so that with 1,000 subjects, a much lower *F* or *r* will be significant than if the sample is 10 or even 100. Statistical significance just tells you that you can be quite sure that there is a systematic relationship between the independent and dependent variables, but it does not tell you how large it is or whether it is practically important. Effect size measures, which are more independent of sample size, tell you how strong the relationship is and thus give you some indication of its importance.

The **profile plots** of cell means (which follow the table of between-subjects effects) help us to visualize the nature of a significant interaction when one exists. When the lines on the profile plot are parallel, there is not a significant interaction. Note that we requested that the separate lines represent the two academic tracks because we felt that this would make a significant interaction easier to interpret than if the two lines represented predominant level of grades. However, really, either independent variable could have been represented in either part of the graph.

Start with the research question.

Remember to always check for an interaction and report that if it is statistically significant.

Example of How to Write About Output 7.9

Results

To assess whether students' higher versus lower math grades and regular versus fast academic track each statistically significantly predict math achievement, and if these relations of math grades to math achievement depend on whether the person is in the fast track or regular track (i.e., on the interaction of math grades with academic track), a two-way ANOVA was conducted. Table 7.6 shows the means and standard deviations for math achievement for the two academic tracks and for the two math grades groups. Table 7.7 shows that there was not a significant interaction between academic track and math grades on math achievement ($p = .563$). There was, however, a statistically significant main effect of academic track on math achievement, $F(1, 71) = 13.87$, $p < .001$. Eta for academic track was .40, which, according to Cohen (1988), is a large effect. Furthermore, there was a statistically significant main effect of math grades on math achievement, $F(1, 71) = 14.77$, $p < .001$. Eta for math grades was about .41, also a medium to large effect.

Be sure to report all main effects that are statistically significant when there is not a statistically significant interaction. If the interaction is significant, the interaction is the principal "effect" to interpret.

Table 7.6
Means, Standard Deviations, and n for Math Achievement as a Function of Academic Track and Math Grades

Math Grade	Fast Track			Regular Track			Total	
	n	*M*	*SD*	*n*	*M*	*SD*	*M*	*SD*
Less A–B	24	12.88	5.73	20	8.33	5.33	10.81	5.94
Most A–B	10	19.27	4.17	21	13.05	7.17	15.05	6.94
Total	34	14.76	6.03	41	10.75	6.70	12.56	6.67

Table 7.7
Analysis of Variance for Math Achievement as a Function of Academic Track and Math Grades

Variable and source	df	MS	F	p	η^2
Math grades	1	515.46	14.77	< .001	.172
Academic track	1	483.93	13.87	< .001	.163
Math grades × academic track	1	11.76	.34	.563	.005
Error	71	34.90			

Interpretation Questions

7.1. (a) Under what conditions would you use a one-sample *t* test? (b) Provide another possible example of its use from the HSB data.

7.2. In Output 7.2: (a) Are the *variances* equal or significantly different for the three dependent variables? (b) List the appropriate *t*, *df*, and *p* (significance level) for each *t* test as you would in an article. (c) Which *t* tests are statistically significant? (d) Write sentences interpreting the academic track difference between the means of *grades in high school* and also *visualization*. (e) Interpret the 95% confidence interval for these two variables. (f) Comment on the effect sizes. Why is Cohen's d selected instead of Hedges' correction or Glass's delta?

7.3. (a) Compare the results of Outputs 7.2 and 7.3. (b) When would you use the Mann-Whitney *U* test?

7.4. In Output 7.2 why do you report Cohen's d and not another measure of effect size?

7.5. In Output 7.4: (a) What does the paired samples <u>correlation</u> for mother's and father's education mean? (b) Interpret/explain the results for the *t* test. (c) Explain how the correlation and the *t* test differ in what information they provide. (d) Describe the results if the *r* was .90 and the *t* was zero. (e) What if *r* was zero and *t* was 5.0?

7.6. (a) Compare the results of Output 7.4 with Output 7.5. (b) When would you use the Wilcoxon test?

7.7. In Output 7.6: (a) Describe the *F*, *df*, and *p* values for each dependent variable as you would in an article. (b) Describe the results in nontechnical terms for visualization and grades. Use the group means in your description.

7.8. In Outputs 7.7 a and b, what pairs of means were significantly different?

7.9. In Output 7.8, interpret the meaning of the sig. values for mathematics achievement and competence. What would you conclude, based on this information, about differences between groups on each of these variables?

7.10. Compare Outputs 7.6 and 7.8 with regard to mathematics achievement. What are the most important differences and similarities?

7.11. In Output 7.9: (a) Is the interaction significant? (b) Examine the profile plot of the cell means that illustrates the interaction. Describe it in words. (c) Is the main effect of academic track significant? Interpret the eta squared. (d) How would you describe the "effect" of math grades? (e) Why did we put the word effect in quotes? (f) Under what conditions would focusing on the main effects be misleading?

Extra SPSS Problems

Using the CollegeStudentData.sav file, do the following problems. Print your outputs after typing your interpretations on them. Please circle the key parts of the output that you use for your interpretation.

7.1. Is there a significant difference between the academic tracks on average student height? Explain. Provide a full interpretation of the results.

7.2. Is there a difference between the number of hours students study and the hours they work? Also, is there an association between the two?

7.3. Write another question that can be answered from the data using a paired sample *t* test. Run the *t* test and provide a full interpretation.

7.4. Are there differences between fast track and regular track students in regard to the average number of hours they (a) study, (b) work, and (c) watch TV? Hours of study is quite skewed, so compute an appropriate nonparametric statistic.

7.5. Identify an example of a variable measured at the scale/normally distributed level for which there is a statistically significant overall difference (*F*) between the three marital status groups. Complete the analysis and interpret the results. Do appropriate post hoc tests.

7.6. Use the Kruskal-Wallis test, with Mann-Whitney post hoc follow-up tests if needed, to run the same problem as 7.1. Compare the results.

7.7. Do students' heights differ depending on academic track and marital status, and do academic track and marital status interact? Run the appropriate analysis and interpret the results.

7.8. Do academic track and having children interact and does either seem to affect current GPA?

7.9. Are there differences between the age groups in regard to the average number of hours they (a) study, (b) work, and (c) watch TV?

8.0. Given the output in 7.9, explain the effect sizes for (a) study, (b) work, and (c) watch TV.

Appendix A
Getting Started and Other Useful SPSS Procedures

An important step when using this book will be to obtain the program. Once you have the program, you will need to download the data we use throughout this book and learn the basics of using the SPSS program. This section will cover copying data files, opening the SPSS application, working with the SPSS syntax, editing your output, importing data and exporting results, standardizing variables, selecting cases, splitting and merging files.

Copy the Data Files From the Website

The datasets are available to **download** from **www.routledge.com/cw/morgan**. They are contained in the **Datasets** (datasets.zip) file under the **Student Resources** tab. The ZIP file contains:

hsbdata.sav (12KB)	**DataFastTrack.sav (11KB)**
CollegeStudentData.sav (4KB)	**DataRegularTrack.sav (11KB)**
ChapterFourData.sav (6KB)	
AlternativehsbdataB.sav (14KB) (to use if you skip Chapter 2 or make an error doing it)	

Note: you may not see the file extension (.sav) depending on your computer setup.

Download these files to a working folder on your computer, personal flash drive, or network drive.

Open and Start the SPSS Application

Begin at the **Start** button (bottom left of the Windows Desktop).

If there is no icon, click **Start** → **IBM SPSS Statistics** → **IBM SPSS Statistics** (see Figure A.1). If **IBM SPSS** is not listed, it will need to be installed on your computer. It is not part of the Microsoft Windows package or the website for this book and must be purchased/rented and loaded separately.

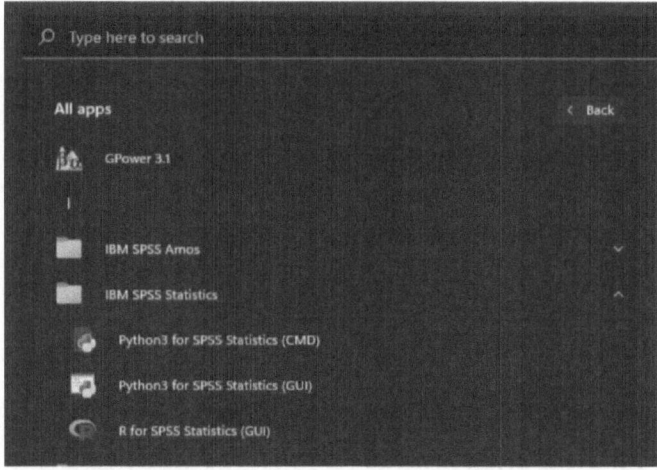

Figure A.1 Start menu and SPSS icon.

After you start the program, you will see the **Welcome to IBM SPSS Statistics** startup screen. There is a What's New channel and links to Help & Support, Tutorials, and Community (see Figure A.2). You can explore these at your leisure to learn more about what IBM SPSS can do for you. However, we want to focus on the **New Files:** on that window. Note that you can also select to view **Restore Points**, **Recent Files**, or **Sample Files**.

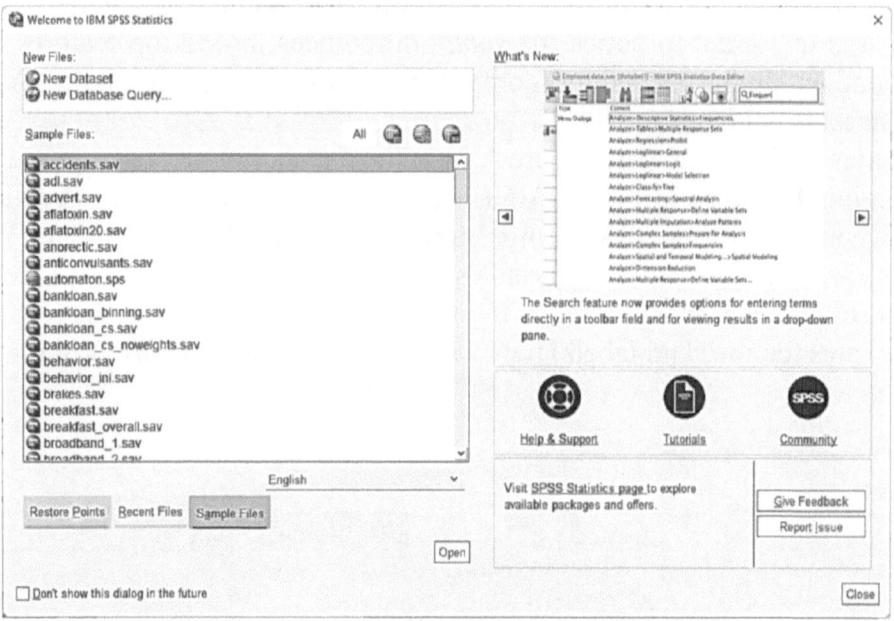

Figure A.2 Startup window.

To open a file:

- If you would like to create a new data file by typing in data or importing it from Excel or another program, under **New Files:** click **New Dataset** → **Open**, *or*
- If you have already created a file, had a problem with SPSS and need to restore a file, or would like to view sample data files under **Restore Points:** select the tab (**Restore Points**, **Recent Files**, or **Sample Files**) → **Open** *or*
- Click the **Close** button, which will bring up a new blank SPSS desktop screen, called the **IBM SPSS Statistics Data Editor**.

    - Note: Files can also be opened from this program's screen by using the **File** menu item.

You should now see the **IBM SPSS Statistics Data Editor** screen, there are two tabs at the bottom left side of the screen; the **Data View** tab and the **Variable View** tab (see Figure A.3).

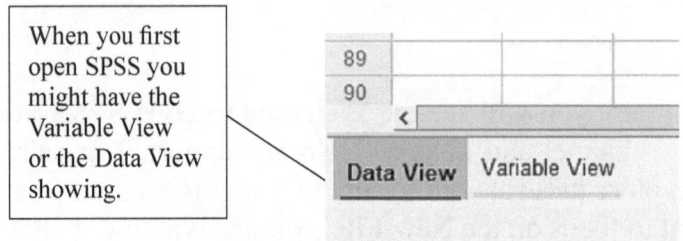

> When you first open SPSS you might have the Variable View or the Data View showing.

Figure A.3 View tabs.

Although the toolbar at the top of the data editor screen is the same for both the Variable and Data View screens, it is important to notice the *subtle* differences in desktop features between these two screens found within the data editor (compare Figs. A.4 and A.5). In Figures A.4 and A.5, the hsbdata.sav dataset is being used.

Select the **Variable View** tab (see Figure A.3) in the data editor screen to produce Figure A.4.

Notice the column headers in Figure A.4 (i.e., Name, Type, Width, etc.). You create (defines and labels) new variables using the **Variable View** (see Chapter 2). As the name "Variable View" implies, in this view, you see names of variables as the data on each row. Notice the column headers are those in Figure A.4, which refer to the variables, not the participants (e.g., Name, Type, Width). One creates (defines and labels) new variables using the **Variable View**.

hsbdata.sav [DataSet1] - IBM SPSS Statistics Data Editor

File   Edit   View   Data   Transform   Analyze   Graphs   Utilities   Extensions   Window   Help

Q Search application

	Name	Type	Width	Decimals	Label	Values	Missing	Columns	Align	Measure	Role
1	acadtrac	Numeric	1	0	academic track	{0, fast track...	None	8	Right	Nominal	Input
2	faed	Numeric	2	0	father's educati...	{2, < h.s. gr...	None	8	Right	Ordinal	Input
3	maed	Numeric	2	0	mother's educ...	{2, < h.s.}	None	8	Right	Ordinal	Input
4	alg1	Numeric	1	0	algebra 1 in h.s.	{0, not take...	None	8	Right	Nominal	Both
5	alg2	Numeric	1	0	algebra 2 in h.s.	{0, not take...	None	8	Right	Nominal	Both
6	geo	Numeric	1	0	geometry in h.s.	{0, not take...	None	8	Right	Nominal	Both
7	trig	Numeric	1	0	trigonometry in ...	{0, not take...	None	8	Right	Nominal	Both

Figure A.4 IBM SPSS Statistics Data Editor: Variable View.

- Click on the **Data View** tab in the data editor to produce Figure A.5.

Notice the column headers change to **var** or to the names of your variables if you have already entered them (see Figure A.5). One enters (inputs) data using the **Data View**. Now, instead of each row referring to a variable, each row refers to data for a particular participant. Notice the column headers change to **var** or to the names of your variables if you have already entered them. So each row shows what score a particular participant has on each of the variables, labeled at the top of the columns. One enters (inputs) data using the **Data View**.

	acadtrac	faed	maed	alg1	alg2	geo	trig	calc	mathgr
1	1	10	10	0	0	0	0	0	0
2	1	2	2	0	0	0	0	0	0
3	1	2	2	0	0	0	0	0	1
4	0	3	3	1	0	0	0	0	0
5	1		3	0	0	0	0	0	0

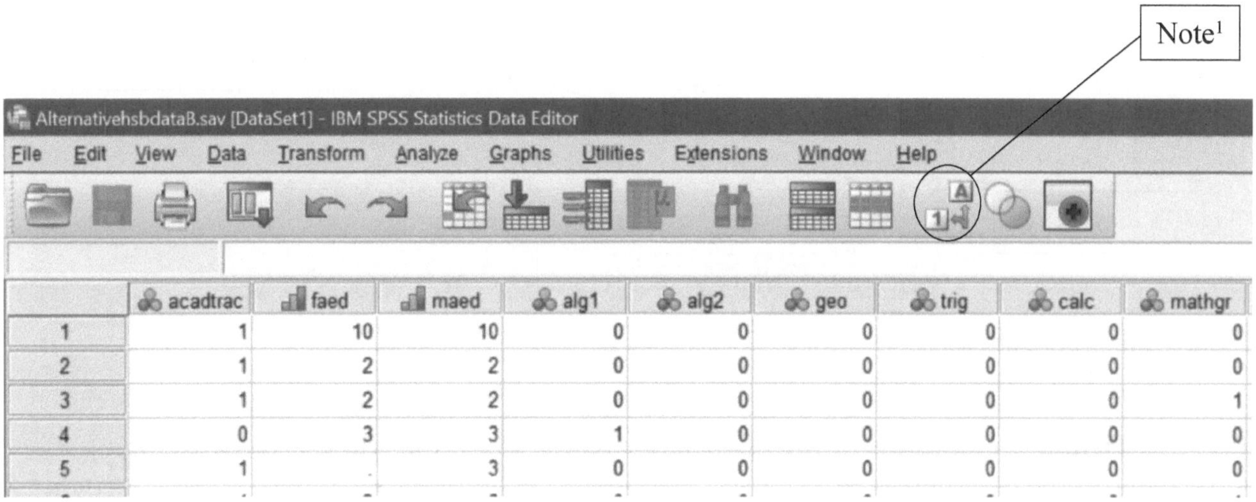

Figure A.5 SPSS data editor: Data View.

Note:

1 If the values for *academic track* are shown as regular track or fast track, the value labels rather than the numerals are being displayed. In that case, click on the circled symbol to change the format to show only the numeric values for each variable.

Set Your Computer to Print the SPSS Syntax (Log)

There are two ways to compute statistics using SPSS: working through the window prompts and syntax. In this book, we focus on using windows, yet knowing how to use syntax can be very helpful. Each time SPSS computes a statistics or graph, syntax is generated. This syntax can be automatically printed on the output. **If you do not see syntax on our output, you can change the settings.** To do so, set your computer using the following:

- Click on **Edit → Options**.
- Click on the **Viewer** tab near the top left of the **Options** window to get Figure A.6 (see the circled tab in Figure A.6).

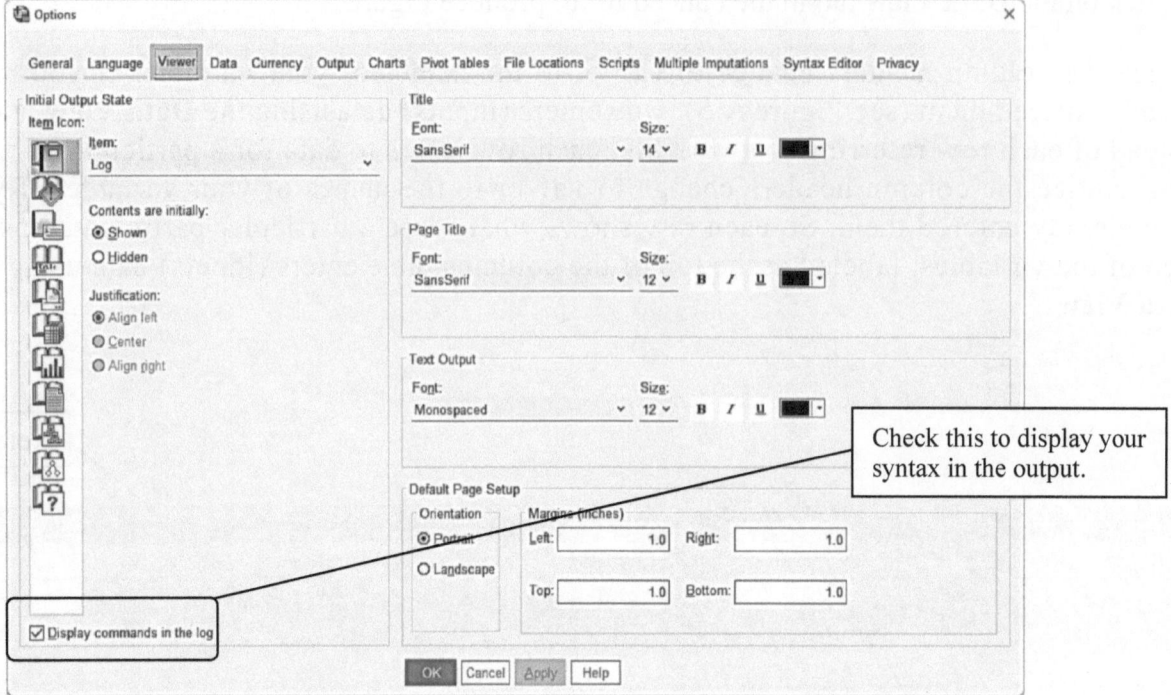

Figure A.6 Edit: Options.

- Check **Display commands in the log** near the lower left of the window (see rectangle).
- Leave the other defaults as is.
- Click on **OK**. Doing this will always print the syntax on your output on this computer. If you use another computer, you may have to repeat the process.

Save Syntax and Later Use It to Rerun Statistics

To save and later use syntax to rerun statistics, you can click on "paste" instead of "OK" before you run the output. This will create a syntax editor window, where you can highlight the part you want to run, then click on the green "play button" (forward arrow). Then, you can save the syntax file.

To save and later use syntax to rerun statistics, you will need to do the following:

- Click on **File → New → Syntax**.
- Copy and paste the syntax from an Output that you have created in SPSS into the syntax editor.
- Click **File → Save as**.
- This will save a **Syntax File (*.sps) →** type file name in dialog box → **Save** (see Figure A.7).

To open and run a saved syntax file, from the menu choose the following:

- With your dataset open click **File → Open → Syntax**.
- Select a syntax file → and click **Open**. Navigate to where you saved the syntax file.
- Once a syntax file is open, select the command (in the example, Figure A.7, its T-TEST).
- From the menu choose **Run → Selection**.

Figure A.7 includes the following syntax for counting frequencies and calculating the mean, median, and mode:

```
FREQUENCIES   VARIABLES =mathach
 /STATISTICS =MEAN MEDIAN MODE
 /ORDER=ANALYSIS
```

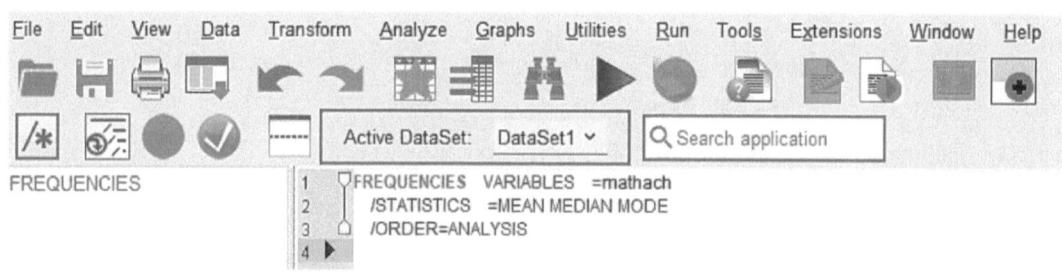

Figure A.7 SPSS Syntax Editor.

Note: This is a simplified version of how to use the syntax file. The saving and using of the syntax file may be different on your computer depending on whether you have a data file open or not. However, it can be very useful if you need to run the same commands several different times.

Editing Your Output

Resize/Rescale to Print

In order for larger tables in the output to fit onto a single printed page, you will need to do the following after computing an output:

- From the IBM SPSS Statistics Viewer, double-click on the table to be resized to open the **Pivot Table Statistics** window (see Figure A.8.). This window provides multiple methods of editing tables and figures.

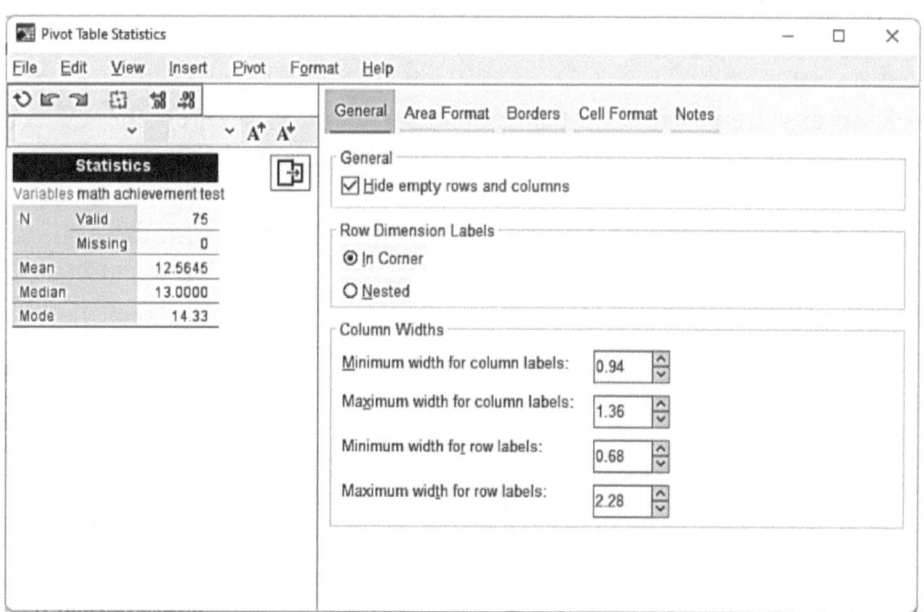

Figure A.8 Pivot Table Statistics.

- Click on File → **Print Properties**. The **Print Options** window appear (see Figure A.9.)
- Check **Rescale wide table to fit page** and/or **Rescale long table to fit page**.
- Click on **OK**.

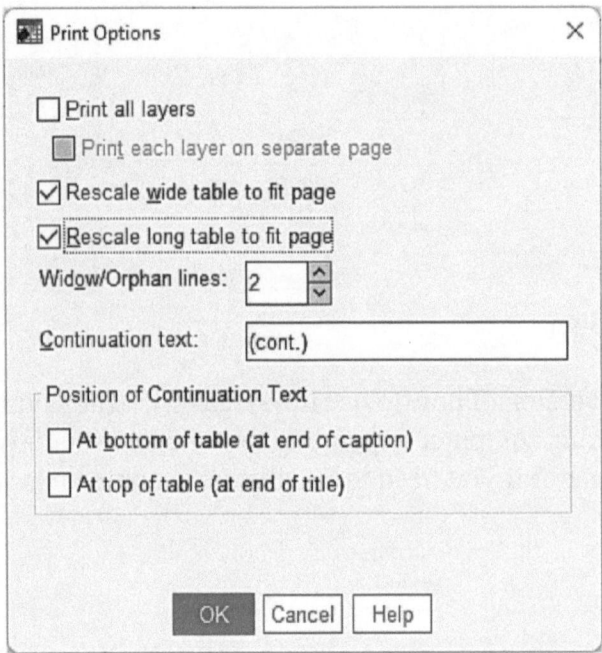

Figure A.9 Print Options.

Editing Tables, Charts, and Text

When using the IBM SPSS Statistics Viewer, editing outputs can be done by double-clicking on any item within the table or chart.

- Double-click on any item within a table or chart. This will open the **Chart Editor** and will allow you to edit the text size, font, color, change the look of a table, or pivot the table.
- Editing of the table or chart can be completed through selecting from the pull-down menus of **Edit**, **Options**, and/or **Elements**. Alternatively, you can click on **View**, and select the toolbars to be shown (see Figure **A.10**).

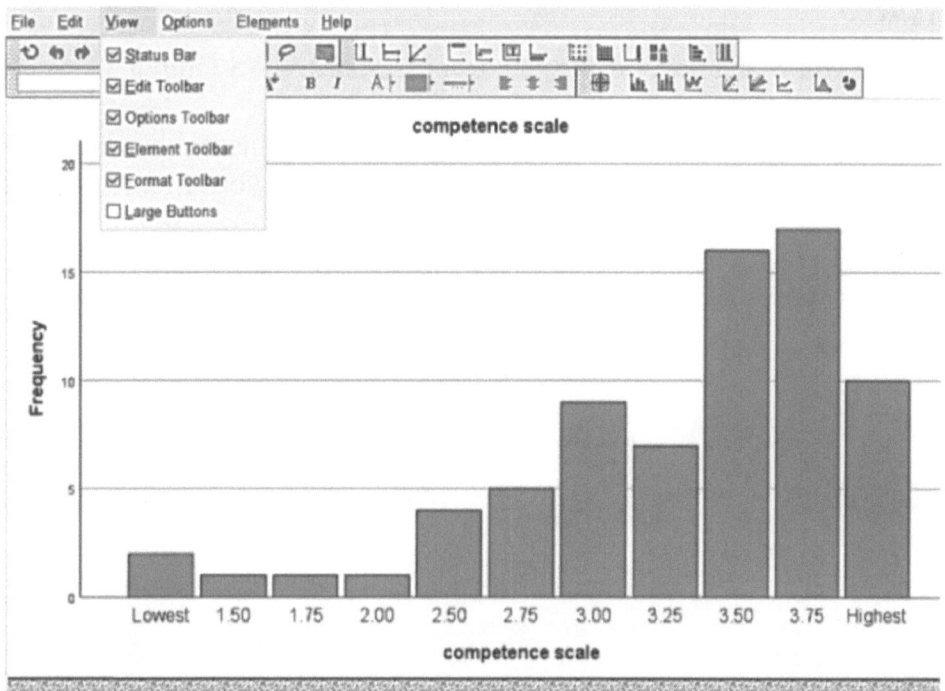

Figure A.10 Chart Editor.

Each toolbar allows you to edit the charts in different ways. If you hover your mouse over the toolbar icons, you will see what each can do.

- The **Formatting Toolbar** allows you to edit text size, font, and color.
- **Table Properties** allows you to customize a table, including text, alignment, shading, footnotes, cell formats, borders, and printing options.
- The **TableLooks** option allows you to change the look of the table.

  - There is a list of predefined styles available.
  - A style can be previewed in the Sample window to the right of the TableLooks dialog box.

- The **Pivot** option lets you move data between columns, rows, and layers.

Importing Data and Exporting Results

SPSS allows you to import data from Microsoft Access or Excel and export results outputs to Microsoft Excel and Word or other file types.

To import from MS Access or Excel:

- From the SPSS Output Viewer menu choose **File → Import Data**. Select the file type you wish to open. A window will open where you can select the file.

To export results:

- When the output you wish to export is open, from the IBM SPSS Output Viewer menu choose **File → Export . . .**
- The **Export Output** window will open (See Figure A.11). Ensure that the box next to **File Name:** includes the drive and folder where you wish to save your file. If it is not listed, then click on **Browse . . .**

Note: You can also export individual elements of the output by right clicking on the element and selecting **Export**.

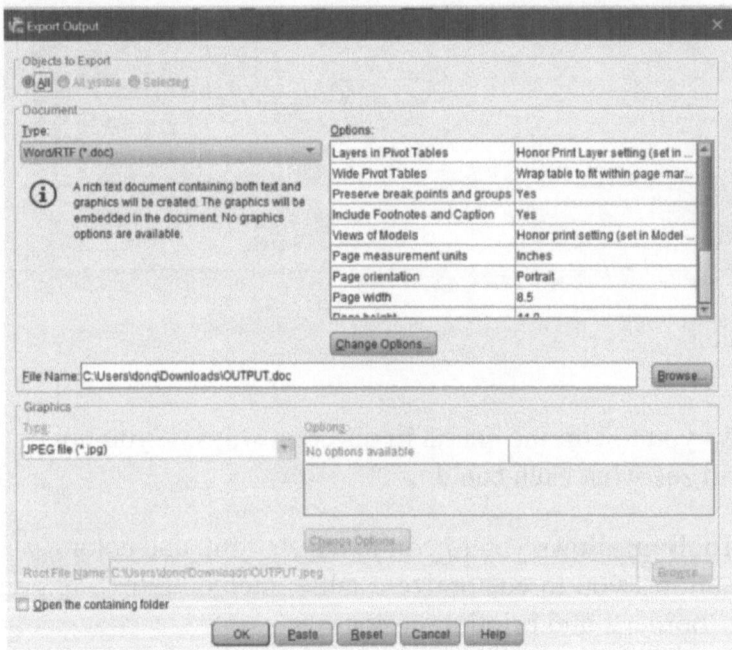

Figure A.11 Export output.

Converting Variables Into Standardized Variables (*z* Scores)

This procedure transforms the data for one variable to a standard score that has a mean of zero and a standard deviation of one. Standardized scores are used when you want to compute a summated scale score made up of variables with quite different means and standard deviations. They are also used to compare apples and oranges, for example, achievement on a math test and an English test. Next we will make the *math achievement* scores into *z* scores.

- Click on **Analyze →Descriptive Statistics → Descriptives . . .**
- Select the variable *math achievement test* (see Figure A.12).
- Click the arrow in the middle of the dialog box to move the variable to the **Variables** box.
- Check the box **Save Standardized Values as Variables**.
- Click **OK.** An output window will appear with the descriptive statistics. The *z* score for each subject will be included as a new variable (*Zmathach*) in the last column of the SPSS Data Editor.

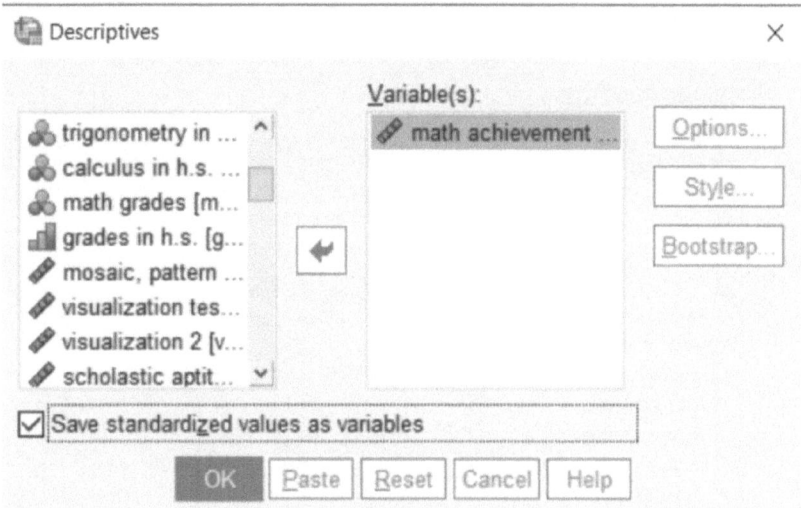

Figure A.12 z scores for math achievement.

Selecting Cases

The select cases command permits the analysis of a specific subset of the data. Once a subset is selected and used for the analysis, the user can either revert to the entire dataset by clicking on reset, or you can delete the unselected cases to create a new data file of the selected cases. If you want to do the same analysis separately on all subsets of data, then **Split File** should be used instead of **Select Cases** (see next section). It is advisable to save your work before deleting cases, just in case you change your mind! To select cases:

- Click on **Data → Select Cases**.
- Choose the method of selecting cases you prefer (See Figure A.13):
  - **If condition is satisfied** (a conditional expression is used to select cases),
  - **Random sample of cases** (cases are selected randomly based on a percent or number of cases),
  - **Based on time or case range** (case selection is based on a range of case numbers or a range of dates/time), or
  - **Use filter variable** (a numeric variable can be used as the filter—any cases with a value other than 0 or missing are selected).
- **Unselected cases** may be **Filtered** (remain in the data file but are excluded in the analysis) or **Deleted** (removed from the working data file and cannot be recovered if the data file is saved after the deletion).
- For example, if you wanted to select only those on the fast track, you would use **If condition is satisfied** and click on the **If . . .** button.
- Then select *academic track* = 0 and click **Continue**.
- Click on **OK**.

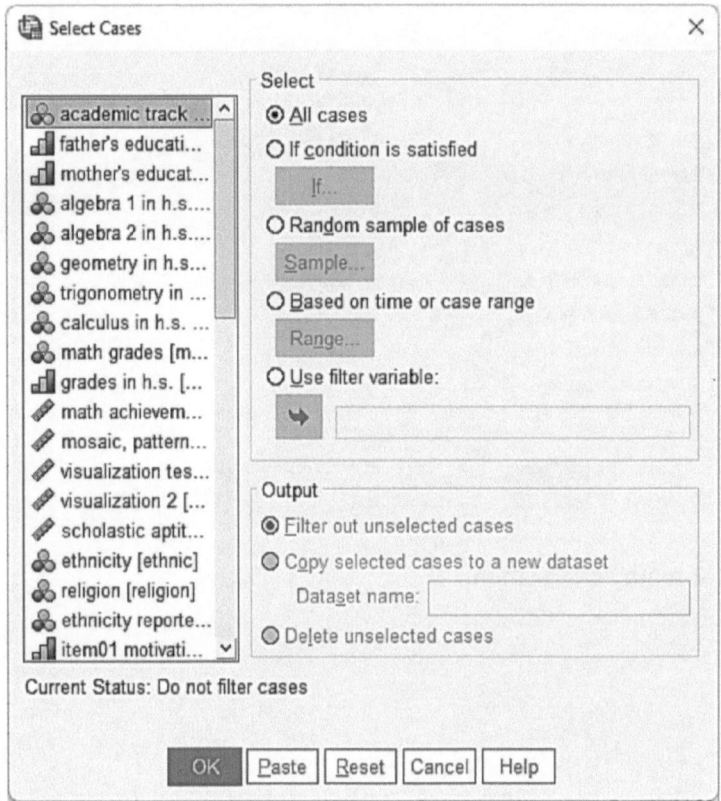

Figure A.13 Select cases.

Splitting Files

Split Files splits the data file into separate groups for analysis based on the values of one or more grouping variables. Data can be displayed for group comparisons, or data can be displayed separately for each group. This is a very useful tool, but <u>be sure to reset **Split File** after doing the analyses you want split, or all further analyses will be split in the same way</u>. In this example, we will split the hsbdata.sav file into two files, one with the data for those in the fast track and one with the data for those in the regular track.

- **Data → Split File**.
- Select the appropriate radio button for the desired display option (**Compare groups** or **Organize output by groups**). Be sure the **Sort the file by grouping variables** is selected. The window should look like Figure A.14.
- Click on **OK**.

Note: Now you can do statistics separately for those in the fast track and those in the regular track.

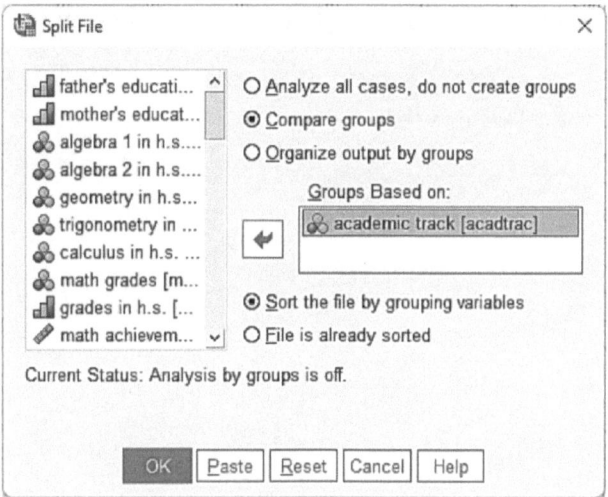

Figure A.14 Split file.

Merging Files

Merge files allows the working data file to be combined with a second data file that contains (a) the *same variables but different cases* or (b) the *same cases but different variables*.

To Add Cases

An example of merging files with the same variables but different cases might be if you had all the participants in the fast academic track for the hsbdata set in one file and all the participants in the regular academic track in another. In order to compare these groups, these two data files need to be merged. It is important that there not be any overlap in participant numbers for the two data files.

- Open both data files you want to merge. In this example, open DataFastTrack.sav and DataRegularTrack.sav. See the access information at the beginning of this chapter.
- In the DataRegularTrack.sav file, click on **Data → Merge Files → Add Cases**.
- The **Add Cases to DataRegularTrack.sav [DataSet2]** window will open (see Figure A.15).
- Highlight **DataFastTrack.sav [DataSet1]** and then click on **Continue**.

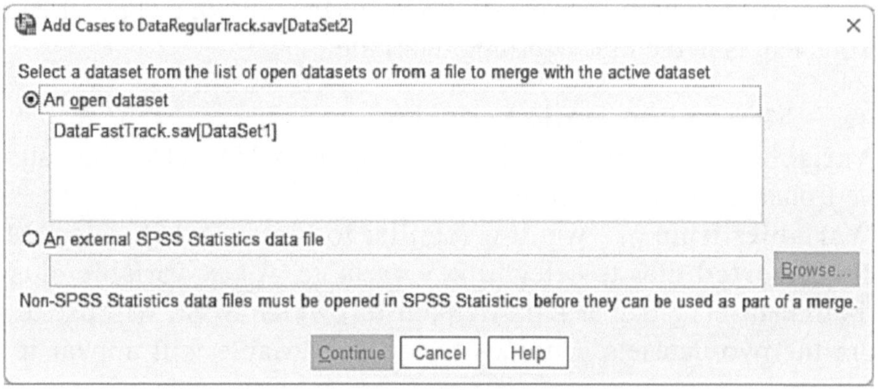

Figure A.15 Add cases to DataRegularTrack.sav [DataSet2].

- The **Add Cases from DataFastTrack.sav [DataSet1]** will open (see Figure A.16).

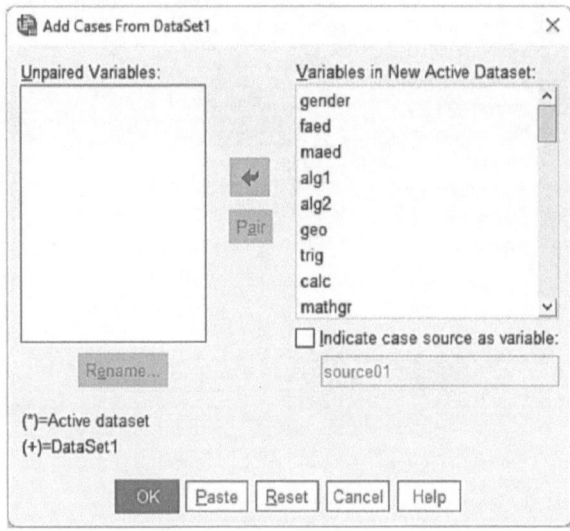

Figure A.16 Add cases from DataFastTrack.sav [DataSet1].

- Click on **OK**. The data from the DataFastTrack.sav file will be added to the DataRegularTrack. sav file.

Using a similar procedure, you can merge two files that have the same, or at least overlapping cases but different variables.

To Add Variables

Before you add <u>variables</u> to a file using this method, you should first make sure that each participant who has data in both datasets is identified by the same ID number in both files. Then, you should use **Sort Cases** to sort each file (by itself), sequentially in ascending order, saving each file once it is sorted. You should make sure that you open the data file <u>first</u> that has the correct values of any variables that exist in both datasets. SPSS will save only one copy of each variable, and that will be the one that is in the first (working data) file.

- Click on **Data → Merge Files → Add Variables**.
- In the **Add Variables to . . .** window (similar to Figure A.15), select the dataset you want to add a variable from and click **Continue**.
- In the **Add Variables from . . .** window (similar to Figure A.16), select **Match cases on key variables in sorted files** to select a key variable. A key variable must be a variable common to both datasets (such as a participant ID) which SPSS will use to match the participants from the two datasets. Choices for a key variable will appear in the **Excluded Variables:** box.

- Click on such a variable and move it into the **Key Variables:** box.
- Click on **OK**.

Case Summaries

A useful report that can be printed with SPSS is called **Case Summaries**. This report lets you see the data for several variables side by side for each participant. You can have the report include all participants or, as we have here, only some. This function could be used to check whether you have computed a transformation correctly. In the partial output here we show how you could visually examine the data for three variables that might be related to *math achievement*.

- In the DataFastTrack.sav file, click on **Analyze → Reports → Case Summaries . . .** The **Summarize Cases** window will open.
- Select the variables *father's education*, *math grades*, *grades in h.s.*, and *math achievement*.
- Click the arrow to move the variables to the **Variables** box.
- Change the "Limit cases to first" box to **10** (see Figure A.17).
- Click on **OK**.

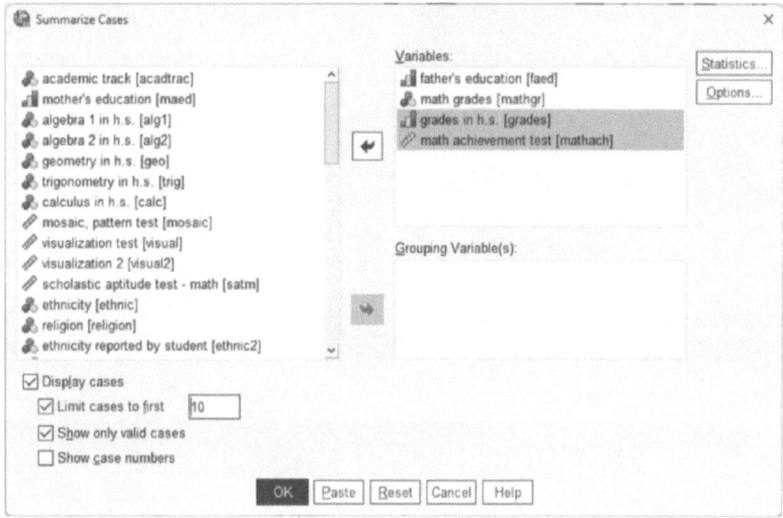

Figure A.17 Summarize cases.

Output A.1: Summarize Cases

Case Processing Summary[a]

	Cases					
	Included		Excluded		Total	
	N	Percent	N	Percent	N	Percent
father's education	9	90.0%	1	10.0%	10	100.0%
math grades	10	100.0%	0	0.0%	10	100.0%
grades in h.s.	10	100.0%	0	0.0%	10	100.0%
math achievement test	10	100.0%	0	0.0%	10	100.0%

a. Limited to first 10 cases.

Case Summaries[a]

	father's education	math grades	grades in h.s.	math achievement test
1	MD/PhD	less A-B	mostly C	9.00
2	< h.s. grad	less A-B	half BC	10.33
3	< h.s. grad	most A-B	mostly B	7.67
4	h.s. grad	less A-B	half CD	5.00
5	.	less A-B	half CD	-1.67
6	h.s. grad	most A-B	half BC	1.00
7	master's	less A-B	mostly B	12.00
8	2 yrs voc	less A-B	mostly C	8.00
9	h.s. grad	most A-B	half AB	13.00
10	coll grad	less A-B	half BC	3.67
Total N	9	10	10	10

a. Limited to first 10 cases.

Marisha Lamont-Manfre

Appendix B
Writing Research Problems and Questions

Frameworks for Stating Research Problems

A common definition of a research problem is that it is a statement that asks what relationship exists between two or more variables. However, most research problems are more complex than this definition implies. The research problem should be a broad statement that covers several more specific research questions to be investigated, perhaps by using summary terms that stand for several variables. Several ways to state the research problem are provided in this appendix. Underlines indicate that you fill in the appropriate name for the variable or group of variables.

Format

One way that you could phrase the problem is as follows: The research problem is to investigate whether (put independent variable 1 or group of variables here) (and independent variable 2, if any, here) (and independent variable 3, if any) are related to (dependent variable 1, here) (and dependent variable 2, if any) in (population here).

Except in a totally descriptive study, there always must be at least two variables (one is usually called the independent variable and one the dependent variable). However, there can be two or more of each, and there often are. In the statement of the problem, in contrast to the research questions/hypotheses, it is desirable to use broad descriptors for groups of similar variables. For example, in the hsb data demographics might cover four variables: geo (geometry in high school or not), mother's and father's education, and ethnicity. Spatial performance might include a mosaic pattern test score and a visualization score. Likewise, grades and mathematics attitudes could each refer to more than one variable. Concepts such as self-esteem or teaching style have several aspects that usually result in more than one variable.

Examples

If your study uses the randomized experimental approach, you could phrase the problem as follows:

1. The research problem is to investigate the effect of a new curriculum on grades, math attitudes, and spatial performance in high school students.

For other studies that compare groups or associate/relate variables, you could phrase the problem as follows:

2. The problem is to investigate whether taking geometry in high school and grades are related to mathematics attitudes and achievement in high school students.

If you have several *independent variables* and want to predict some outcome, you could say:

3. The problem is to investigate the variables that predict or *seem* to influence mathematics achievement.

This latter format is especially useful when the approach is a complex (several independent variables) associational one that will use multiple regression.

Framework for Stating Research Questions/Hypotheses

Although it is okay to phrase a randomized experimental research problem (in the format of the first example earlier) as a "study of the effect of . . . " we think when a study is not a randomized experiment, it is best to phrase your research questions or hypotheses so that they do not appear to imply cause and effect (i.e., as *difference* or *associational* questions/hypotheses and/or as *descriptive* questions). The former are answered with inferential statistics, and descriptive questions are answered with descriptive statistics. There are several reasonable ways to state research questions. In the following, we show one way to state each type of question, which we have found useful and, hopefully, clear for our students.

Descriptive Questions

Basic descriptive questions. Descriptive questions ask about the central tendency, frequency distribution, percentage in each category, variability, or shape of the distribution of a variable. Some descriptive questions are intended to test assumptions. Some questions simply describe the sample demographics; others describe a dependent variable. A few *examples* are as follows:

1. Is mathematics achievement distributed approximately normally?
2. What percentage of participants took geometry?
3. What are the mean, mode, and median of the mathematics achievement scores?

Complex descriptive questions. These questions deal with two or more variables at a time, but do not involve inferential statistics. Cross-tabulations of two categorical variables, factor analysis, and measures of reliability (e.g., Cronbach's alpha) are examples.

Two *examples* are as follows:

1. What is the internal consistency reliability of the pleasure scale items?
2. What are the percentages of those who took geometry and those who did not between the experimental and control groups?

Difference Questions/Hypotheses

Basic Difference Questions

The *format* is:

Are there statistically significant differences between the (<u>insert number</u>) levels of (<u>put the independent variable name here</u>) (<u>you could name the levels here in parentheses</u>) in regard to the average (<u>put the dependent variable name here</u>) scores? Another acceptable format is shown in example 2.

Two *examples* are as follows:

1. Are there statistically significant differences between the three levels (high, medium, and low) of father's education in regard to the average mathematics achievement scores of the students?
2. Is there a statistically significant difference between those who took geometry or not on the visualization score?

Appropriate analyses: One-way ANOVA. A *t* test could be used if there were only two levels of the independent variable, as in example 2.

Complex difference and interaction questions. When you have two categorical independent variables considered together, you will have *three* research questions or hypotheses. There are advantages to considering two or three independent variables at a time. See our chapter on ANOVA for an introduction about how to interpret the *interaction* question. Sample *formats* for a set of three questions answered by *one* two-way ANOVA are as follows:

1. Is there a statistically significant difference between (<u>insert the levels of independent variable 1</u>) in regard to the average (<u>put dependent variable 1 here</u>) scores?
2. Is there a statistically significant difference between (<u>insert the levels of independent variable 2</u>) in regard to the average (<u>dependent variable 1</u>) scores?
3. Is there a statistically significant interaction of (<u>independent **variable 1**</u>) and (<u>independent **variable 2**</u>) in regard to the (**<u>dependent</u> variable 1**)?

(Repeat these three questions, for the second dependent variable, if there is more than one.) An *example* is as follows:

1. Is there a statistically significant difference between students who have high versus low math grades in regard to their average mathematics achievement scores?
2. Is there a statistically significant difference between those who took Algebra I in high school and those who did not in regard to their average math achievement scores?
3. Is there a statistically significant interaction between mathematics grades and taking Algebra I on math achievement?

Note that the first question states the *levels* or categories of the first independent variable; that is, it states the groups that are to be compared (high vs. low math grade students). The second question does the same for the second independent variable; that is, it states the *levels* (Algebra I or not) to be compared. However, the third (interaction) question asks whether the first *variable* itself (mathematics grades) interacts with the second variable (alg1). No mention is made, in the interaction question, of the values or levels or groups.

An appropriate analysis: Factorial ANOVA.

Associational/Relationship Questions/Hypotheses

Basic associational questions. When both variables are ordered and essentially continuous (i.e., have five or more ordered categories), we consider the approach and research question to be associational. There are two main types of basic associational statistics: correlation and regression.

The *format* for a correlation is as follows:

Is there a statistically significant association between (<u>variable 1</u>) and (<u>variable 2</u>)?

In this case, it is arbitrary which variable is independent or antecedent and which is dependent or outcome unless one occurs before the other in time; see example 3. An *example* for a single association or relationship is as follows:

1. Is there a statistically significant association between grades in high school and mathematics achievement?

If there are more than two variables, which is common, and each pair of variables is associated separately, you can have a series of questions asking whether there is an association between *each* variable and every other variable. This would produce a *correlation matrix*.

An *example* that would produce a correlation matrix is as follows:

2. Are there statistically significant associations among the <u>three</u> mathematics attitude scale scores?

Note that what is said to be associated in these questions is the variable itself; no mention is made of the levels or values here.

If one variable is clearly the independent, antecedent, or predictor, you would phrase the question as follows and use *bivariate regression* analyses:

3. How well can we predict math achievement test scores (the dependent variable) from grades in high school (the independent variable)?

Appropriate analyses: Bivariate regression if there is a clear independent or antecedent variable and you want to make a prediction; correlation if there is no clear independent variable.

Complex associational questions. In the associational approach, when two or more *independent* variables are considered together, rather than separately, as in the previous basic format, you get a new kind of question. The *format* can be phrased something like:

How well does the combination of (<u>list the several specific independent variables here</u>) predict (<u>put dependent variable here</u>)?

Two *examples* are as follows:

1. How well does the combination of number of mathematics courses taken, if the students took Algebra I or not, and father's education predict mathematics achievement?
2. How well does a combination of H.S. ACT, GPA, and rank in class predict first year college GPA?

An appropriate analysis: Multiple regression.

The first complex question in the previous section also could be expanded, into a set of questions, to help you understand more fully. This set first asks about the association of each of the predictors (or independent) variables and the dependent (or outcome) variable and then states the complex or combination question as earlier.

For *example*:

1. Is there a statistically significant association between the number of mathematics courses taken and mathematics achievement test scores?
2. Is there a statistically significant association between math grades (high or low) on mathematics achievement?
3. Is there a statistically significant association between father's education and mathematics achievement?
4. How well does the combination of the number of mathematics courses taken, math grades (low or high), and father's education predict mathematics achievement test scores?

Appropriate analysis: The multiple regression output will provide you with the three bivariate, Pearson correlations in a matrix as well as the multiple regression statistics.

Appendix C

Answers to Odd Numbered Interpretation Questions

Chapter 1: Variables, Research Problems, and Questions

1.1. Compare the terms *active independent variable* and *attribute independent variable*. What are the similarities and differences?

An *attribute independent variable* is a characteristic or a "part" of the participants, whether or not the study takes place. It is not manipulated by the researcher. Some common attribute variables, such as ethnicity, gender, and IQ, are not easily changed. Other attributes, such as age, income, job title, personality traits, and attitudes, can change over time. The key characteristic of *attribute independent variables* is that they are *measured, not manipulated*. An *active independent variable* is manipulated or caused to vary systematically. The treatment or condition that the group receives may be determined by the experimenter or sometimes by someone else, often a group (school, clinic, etc.). The researcher or other person actively gives different groups different treatments, and the differences between groups that have different treatments are usually the focus of the study. The most common active variable is when an experimental and comparison group receive different treatments (curricula, interventions, etc.). The participants might be randomly assigned to groups or already be in intact groups such as school classes.

1.3. What is the difference between the independent variable and the dependent variable?

In the classic definition of the independent variable, it is the variable that is manipulated; however, in this book and commonly in the field, the term "independent variable" is used more broadly to mean the *presumed cause* of differences in the outcome variable. The scores or values for the dependent variable "depend on" the level of the independent variable. For example, you might have an experiment in which you are testing the effectiveness of a new weight reduction plan, say reduced carbohydrates. The independent variable involves whether or not the participant was on the low carbohydrate diet. One group (one level of the independent variable) is given and follows a low carbohydrate diet, whereas the other group (the other level of the independent variable) eats their normal diet. Then to determine if the low carbohydrate system works, the researcher might make weight measurements before the low carbohydrate diet is initiated and again following a certain period of time (during which one group was on the low carbohydrate diet and one was not). The weight measurements serve as the dependent variable. Thus, the researcher is hoping that the low carbohydrate treatment (the independent variable) affects a change in weight (the dependent variable).

1.5. Write a research question *and* a corresponding hypothesis regarding variables of interest to you but not in the HSB dataset. Is it an associational, difference, or descriptive question?

Of course the answers to this question will vary greatly. An associational question will most likely involve the relation between two normally distributed variables, for example, "Is there an association between IQ scores and SAT scores for high school seniors?" Difference questions usually compare two to four groups on some outcome variable. For example, "Are there differences between three different weight loss programs in regard to the average weight loss?" Although it should be the easiest, often students misunderstand descriptive questions. Descriptive questions do NOT try to make inferences about the larger population, and group differences are associations that exist generally. They are limited to trying to understand the distributional characteristics of the data for the specific group of people who were studied. You might ask, "What is the average weight loss of all participants in the study?" Or "What is the average age of all participants in the study?" Descriptive questions are not answered with inferential statistics such as a *t* test.

1.7. Why is it important to classify dichotomous differently than nominal?

Classifying dichotomous variables differently from nominal variables is important because their distinct characteristics influence how they are analyzed and interpreted in research. For example, dichotomous variables have only two categories, enabling analysis using statistical methods like Pearson's r, Spearman rho, multiple regression, and more rarely, logistic regression or binary outcomes in predictive models, and results often boils down to "yes vs. no." Nominal variables have three or more categories without an order. An example of a nominal or categorical variable is ethnicity or religious affiliation when using three or more levels.

Chapter 2: Getting Data Ready for Analysis and Understanding It: Data Collection, Coding, and Description

2.1. Using Outputs 2.5a and 2.5b: (a) What is the mean *visualization test* score? (b) What is the range for grades in h.s.? (c) What is the minimal score for *mosaic pattern* test? How does this compare to the value of that variable as indicated in Chapter 1? Why could the minimal be a negative number?

   (a) The mean of visualization test score is 5.26.
   (b) The range for grades in h.s. is 6.
   (c) The minimal score for *mosaic pattern* test is –4.0. This may seem like an error. However, if you check the codebook, you will see visualization scores go from –4 to 16. The –4 score verifies at least one person scored the lowest possible score.

2.3. Using Output 2.7: (a) Can you interpret the means? Explain. (b) How many participants are there altogether? (c) How many have completed data (nothing missing)? (d) What percent took *algebra 1*?

   (a) No. Because the mean is not meaningful for nominal variables.
   (b) 75
   (c) 75
   (d) 78.7%

2.5. (a) Compare and contrast nominal, dichotomous, ordinal, and normal variables. (b) In social science research, why isn't it important to distinguish between interval and ratio variables?

(a) *Table AB.1*

	Nominal	*Dichotomous*	*Ordinal*	*Normal (scale)*
Characteristics	Categories without an order	Two-category variable	Ordered categories, but intervals are not equal	Continuous variable with equal intervals
Examples	Eye color Types of music	Yes/No High vs. low	Education levels Socioeconomic status	Age Temperature

(b) The distinction between interval and ratio variables is not important in social science research because it does not influence how the data is analyzed or interpreted in most studies. In contrast, experimental sciences often need a clear distinction, as the presence of an absolute zero point in ratio variables is crucial for measurement and interpretation.

2.7. (a) How do z scores relate to the normal curve? (b) How would you interpret a z score of −3.0? (c) What percentage of scores is between a z of −2 and a z of +2? Why is this important?

(a) Z scores represent the number of standard deviations a score is from the mean of a distribution. It allows to standardize scores, enabling comparisons across different distributions.

(b) A z score of −3.0 means the score is 3 standard deviations below the mean of the distribution.

(c) In a normal distribution, approximately 95% of the scores fall between $z = -2$ and $z = +2$. This range represents two standard deviations above and below the mean. This range captures the majority of the data in a normal distribution, and scores outside this range are considered outliers.

Chapter 3: Selecting and Interpreting Inferential Statistics

3.1. Compare and contrast a between-groups design and a within-subjects design.

In a *between-groups design*, different people get different treatments or have different levels of the independent variable. In other words, each level of the independent variable indicates a particular treatment that participants in that group received or a particular attribute participants share, and participants are classified into groups based on these differences. For example, you might want to see if one 8th grade math curriculum is more effective than another. Each level of the independent variable in this between-groups design would involve a different set of students.

In a *within-subjects design*, the same people get multiple treatments or are measured on the same variable at different times (or related/matched people get the same treatment or are measured on the same variable). This design is sometimes referred to as a repeated measures design. The most common example of this design is a pretest posttest design. All students take the pretest and the same students take the posttest. Another example is when a group of people are being monitored over time. Twenty people might enter an exercise program and their blood pressure and cholesterol levels might be measured each week for 10 weeks. The independent variable in a within-subjects design is sometimes referred to as "time" and is used to assess how the dependent variable changes as a function of time (i.e., differences across different levels of the "time" variable in the dependent variable). For example, "Was there a significant change in the cholesterol levels and blood pressure levels over the ten-week period?" is a within-subjects design question. It is also a within-subjects design if pairs of subjects are matched and then compared because the subjects are systematically related to each other and do not meet the assumptions of a between-subjects design. It is also a within-subjects design if there is a family link between participants. For example, is there a difference between students' height and students' same sex parent's height?

3.3. Provide an example of a study, including identifying the variables, level of measurement, and hypotheses, for which a researcher could appropriately choose two different statistics to examine the relations between the same variables. Explain your answer.

Answers will vary. We have presented an example to assist in your understanding:

Hypothesis: More guilt-prone adults are more likely to help out a needy child than are less guilt-prone adults.
IV: guilt proneness on an ordered 5-point scale (20+ adults at each level)
DV: helping behavior (normally distributed), measured by how much of the participant's prize money for participating in a study is donated to "a needy child" (jar with poster of child and a sign requesting help).

One could use ANOVA to compare the five groups, with each level of guilt proneness creating a separate group of participants. Post hoc tests would allow one to determine whether each pair of guilt levels differ; the hypothesis predicts that high (e.g., levels 4 and 5) guilt-prone adults would help more than low (e.g., levels 1 and 2) guilt-prone persons. Because both variables have five or more levels, one could also use correlation to examine this if one thought that there should be a linear relation between guilt proneness and altruism.

3.5. Interpret the following related to effect size:

(a)	$d = .25$	Small or smaller than typical
(b)	$r = .35$	Medium or typical
(c)	$R = .53$	Large or larger than typical
(d)	$r = .13$	Small or smaller than typical
(e)	$d = 1.15$	Much larger than typical
(f)	eta $= .38$	Large or larger than typical

3.7. What statistic would you use if you had two independent variables, income group ($< \$10,000$, $\$10,000$–$\$30,000$, $> \$30,000$) and ethnic group (Hispanic, Caucasian, African-American), and one normally distributed dependent variable (self-efficacy at work)?

Factorial ANOVA (this would be a 3×3 factorial ANOVA).

3.9. What statistic would you use if you had three normally distributed (scale) independent variables (weight of participants, age of participants, and height of participants), plus one dichotomous independent variable (academic track) and one dependent variable (positive self-image), which is normally distributed?

Multiple regression.

3.11. Results of a *t* test reported a difference between the stress levels of university students who had children compared to university students who did not have children. Using the Perceived Stress Score (PSS), university students with children averaged 7.2 and the stress level for university students without children was 5.6. The results of the t test were $t = 3.52$, $p = .032$, and $d = .84$. Using the steps in Figure 3.4, write in interpretation of these results.

University students with children were statistically different from university students without children on Perceived Stress Score, $t = 3.52$, $p = .032$. The means indicate university students with children ($M = 7.2$) is significantly higher than the score for university students without children ($M = 5.6$). The difference between the means is 1.6, and the effect size d is .84, which is a large or larger than typical.

Chapter 4: Methods to Provide Evidence for Reliability and Validity

4.1. Write an additional sentence or two describing the disagreements in Output 4.1 that you might include in a detailed research report.

There were six disagreements between the ethnicity reported in the school records and the ethnicity reported by the student. These included one student whose record indicated they were Euro-American, but who self-identified as African-American; three students whose record indicated they were African-American but two self-identified as Euro-American and one as Latino-American; and two students who self-identified as African-American but whose school's record showed they were Latino-American and Asian-American.

These disagreements are important to consider since they may be errors in the school's record, errors in the self-identified data, or the student may self-identify themselves based on different criteria.

4.3. In Output 4.3, which factor explained the most variance? Why is this important?

Factor 1 explained the most variance (21.55%). This is important to consider because if there is one factor that explains most of the variance, then the scale may only have one factor to consider.

4.5. If you had to delete one question from the analysis in Output 4.4a, which one would you delete?

By looking at the Item-Total Statistics table, the last column under "Cronbach's Alpha if Item Deleted," one can see that item 05 reversed would decrease the alpha the most if it were deleted, so you wouldn't want to delete it. Item 11 reversed is the best one to delete if we had to delete one because deleting it would increase Cronbach's alpha from .856 to .869.

Chapter 5: Chi-Square

5.1. In Output 5.1: (a) What do the terms "count" and "expected count" mean? (b) What does the difference between them tell you?

(a) The count is the actual number of subjects in that cell. For example, in this output, 24 males and 20 females had low math grades. The expected count is what you would expect to find in the cell given the marginal totals (the values in the "total" column and row) if there were no relationship between the variables.

(b) If the expected count and the actual count are similar, there is probably not a significant difference between males and females in this dataset. If, however, there is a large difference between expected and actual count, you would expect to find a significant chi-square.

5.3. In Output 5.2: (a) How is the risk ratio calculated? What does it tell you? (b) How is the odds ratio calculated and what does it tell you? (c) How could information about the odds ratio be useful to people wanting to know the practical importance of research results? (d) What are some limitations of the odd ratio as an effective size measure?

(a) The risk ratio for students with low math grades is the percentage of those students who did not take algebra 2 who have low math grades (70%), divided by the percentage of those students who did take algebra 2 who have low math grades (45%). In this case, students who do not take algebra 2 are about 1 ½ times as likely to have low math grades as those who do take algebra 2.

(b) The odds ratio is the ratio of the risk ratios for students with low and with high math grads. It tells you that one is almost three times as likely to get low grades as high grades if one did not take algebra 2.

(c) The odds ratio is useful because it describes the likelihood that people will have a certain outcome given a certain other condition. One can then judge whether or not these odds justify the intervention/treatment.

(d) Unfortunately, there are no agreed upon standards as to what is a large odds ratio.

5.5. In Output 5.4: (a) How do you know which is the appropriate value of eta? (b) Do you think it is high or low? Why? (c) How would you describe the results?

(a) The eta with math courses taken as the dependent variable is the appropriate eta, because we are thinking of gender as predicting math courses taken, rather than the reverse.

(b) It is medium (average) to large, using Cohen's criteria; however, gender does not explain very much variance (.11) in how many math courses are taken.

(c) The results indicate that boys are likely to take more math courses than are girls.

Chapter 6: Correlation and Regression

6.1. Why would we graph scatterplots and regression lines?

The most important reason is to check for violations in the assumptions of correlation and regression. Both the Pearson correlation and regression statistic assume a linear relationship. Viewing the scatterplot lines allows the researcher to check to see if there are marked violations of linearity (e.g., the data may be curvilinear). In regression, there may be better fitting lines such as a quadratic (one bend) or cubic (two bends) that would explain the data more accurately. Graphing the data also allows you to easily see outliers.

6.3. In Output 6.3, how many of the Pearson correlation coefficients are significant? Write an interpretation of (a) one of the significant and (b) one of the nonsignificant correlations in Output 6.3. Include whether or not the correlation is significant, your decision about the null hypothesis, *and* a sentence or two describing the correlations in nontechnical terms. Include comments related to the sign and to the effect size.

There are five significant correlations: (1) visualization with scholastic aptitude test—math, (2) visualization with math achievement test, (3) scholastic aptitude test—math and grades in high school, (4) scholastic aptitude test—math with math achievement, and (5) grades in high school with math achievement. There are several possible answers to the rest of this question; we present two examples.

(a) s+There was a significant positive association between scores on a math achievement test and grades in high school ($r_{(73)}$ = .504, $p < .001$). In general, those who scored high on the math achievement test also had higher grades in high school. Likewise, those who did not score well on the math achievement test did not do as well on their high school grades. The effect size ($r = .50$) was larger than typical. The null hypothesis stating that there was no relationship can be rejected.

(b) There was not a significant relationship between the visualization test scores and grades in high school ($r_{(73)}$ = .127, $p = .279$). Although this could be said to be a very small positive effect size, the direction of the relationship and effect size should not be discussed because the magnitude of this correlation is so low (and nonsignificant) that it can be viewed as a chance deviation from zero. There is little evidence from this dataset to support a relationship between these two variables. The null hypothesis is not rejected.

6.5. In Output 6.5, what do the standardized regression weights or coefficients tell you about the ability of the predictors to predict the dependent variable?

The standardized coefficients allow you to compare the amount that each variable contributes to predicting the outcome (like *math achievement* in Output 6.5) when all variables are used as predictors. Standardizing predictors put them all on the same scale, so you can identify which predictors most strongly predict the dependent variable.

Chapter 7: Comparing Groups with *t* Tests, Analysis of Variance (ANOVA), and Similar Nonparametric Tests

7.1. (a) Under what conditions would you use a one-sample *t* test? (b) Provide another possible example of its use from the HSB data.

(a) It is not uncommon to want to compare the mean of a variable in your dataset to another mean for which you do not have the individuals' scores. One example of this is comparing a sample with the national norm. You could also compare the mean of your sample to that from a different study. For example, you might want to replicate a study involving GPA, and ask how the GPA in your study this year compares to the GPA in the replicated study of ten years ago, but you only have the mean GPA (not the raw data) from that study.

(b) We could compare the mean in the HSB dataset with national norms for the visualization test, the mosaic pattern test, or the math achievement test. Comparing our data with national norms could help us justify that the HSB dataset is similar to all students or tell us that there is a significant difference between our HSB data and national norms.

7.3. (a) Compare the results of Outputs 7.2 and 7.3. (b) When would you use the Mann-Whitney *U* test?

(a) Note that, although the two tests are based on different assumptions, the significance levels (i.e., *p* values) and results were similar for the *t* test and the M -W that used the same variables. Males and females were significantly different on math achievement ($p = .009$ and .010, respectively) and on visualization scores ($p = .020$ and .040), but there was not a significant difference between males and females on grades in high school ($p = .369$ and .413).

(b) The Mann-Whitney *U* is used to compare two groups, as is a *t* test, but you should use the M-W when you have ordinal (not normally distributed) dependent variable data. The M-W can also be used when the assumption of equal group variances is violated.

7.5. In Output 7.4: (a) What does the paired samples <u>correlation</u> for mother's and father's education mean? (b) Interpret/explain the results for the *t* test. (c) Explain how the correlation and the *t* test differ in what information they provide. (d) Describe the results if the *r* was .90 and the *t* was zero. (e) What if *r* was zero and *t* was 5.0?

(a) The paired samples correlation means a positive relationship between the two variables: mother's and father's education. Specifically, as the education level of mother increases, the education level of the father tends to increase as well. This correlation is statistically significant ($p< .001$).

(b) The *t* test result shows a statistically significant difference between fathers' and mothers' education, $t (73) = 2.40$, $p = .02$. Inspection of mother's and father's education means indicates that the average father's education is significantly higher than mother's education and the difference is .59.

(c) The correlation measures the strength and direction of the relationship between fathers' and mothers' education. It does not assess differences in their averages, but instead how well one parent's education predicts the other's. The t test suggests there is a significant average difference in father's and mother's education.

(d) A correlation of 0.90 indicates a large and positive relationship between fathers and mothers' education. Specifically, as father's education increases, mother's education increases, too. A *t* test result of zero means there is no significant difference between mother's and father's education.

(e) A correlation of 0 indicates no relationship between fathers' and mothers' education. Specifically, father's education cannot predict mother's education. $T = 5.0$ indicates a statistically significant mean difference between fathers' and mothers' education.

7.7.   In Output 7.6: (a) Describe the *F*, *df*, and *p* value for each dependent variable as you would in an article. (b) Describe the results in nontechnical terms for visualization and grades. Use the groups means in your description.

(a)   Grades in h.s: $F(2,70) = 4.09$, $p = .021$. This result indicates that there is a statistically significant difference in grades in h.s. among the three groups. Visualization Test: $F(2,70) = 0.763$, $p = .47$. This result shows that there is no statistically significant difference in visualization test scores among the three groups. Math Achievement Test: $F(2,70) = 7.88$, $p = .001$. This result indicates that there is a statistically significant difference in math achievement test scores among the three groups.

(b)   The visualization test scores were very similar across the three groups: HS Grad or Less: $M = 4.677$, $M = 4.677$, Some College: $M = 6.015$, $M = 6.015$, BS or More: $M = 5.486$, $M = 5.486$. The differences between these group means were not statistically significant ($p = .47$). The average grades in high school varied slightly between the groups: HS Grad or Less: $M = 5.34$, $M = 5.34$, Some College: $M = 5.56$, $M = 5.56$, BS or More: $M = 6.35$, $M = 6.35$. The difference in these means was statistically significant ($p = .02$).

7.9.   In Output 7.8, interpret the meaning of the sig. value for mathematics achievement and competence. What would you conclude based on this information about differences between groups on each of these variables?

Students whose fathers had some college education had a significantly higher math achievement mean rank (43.78) than those whose fathers were high school graduates or less (28.43). This difference was statistically significant ($z = -2.43$, $p = .015$), and the effect size was medium to large ($r = .33$). Also students whose fathers had a bachelor's degree or more had a significantly higher mean rank (48.42) compared to those whose fathers were high school graduates or less (28.43). There was a significant difference ($z = -3.36$, $p = .001$), and the effect size was medium to large ($r = .46$). Further, there was no statistically significant difference in math achievement between students whose fathers had some college education and those whose fathers had a bachelor's degree or more ($z = -0.65$, $p = .52$).

What could be concluded is mathematics achievement scores were significantly lower for students whose fathers were high school graduates or less compared to those whose fathers had some college or a bachelor's degree or more. However, there was no significant difference in math achievement between students whose fathers had some college and those whose fathers had a bachelor's degree or more.

7.11.   In Output 7.9: (a) Is the interaction significant? (b) Examine the profile plot of the cell means that illustrates the interaction. Describe it in words. (c) Is the main effect of academic tracking significant? Interpret the eta squared. (d) How would you describe the "effect" of math grades? (e) Why did we put the word effect in quotes? (f) Under what conditions would focusing on the main effects be misleading?

(a)   No.

(b)   The profile plot indicates that while both fast track and regular track students perform better in math achievement as their math grades improve. Fast track students outperform regular track students.

(c) Yes, $F(1,71) = 13.866$, $p < .001$. The partial eta squared value of .163, suggesting that 16.3% of the variance in math achievement scores can be attributed to academic tracking, after controlling for other variables in the mode.

(d) The effect of math grades on math achievement is statistically significant, $F(1,71) = 14.77$, $p < .001$. 17.2% ($\eta^2 = .172$) of the variance in math achievement scores is explained by differences in math grades.

(e) This is not a causal experimental design, so math achievement is not caused by or were the effect of math grades and academic track.

(f) Focusing solely on main effects is misleading when interaction effects or conditional relationships influence the dependent variable.

Xia Xue

Appendix D
Glossary

Active (Independent Variable): A variable, such as a workshop, new curriculum, or other intervention, at least one level of which is given to a group of participants, within a specified period of time during the study.

Alpha: The level set by the researcher to test null hypothesis, usually this is set at .05 but can be set at different levels for research with greater risk involved if making a type I error (e.g., medical research).

Approximately Normally Distributed: When there are scores (usually interval or scale) with few scores on the ends and most scores in the middle. We define a normal distribution as a set of scores with a skewness less than the absolute value of one.

Associational Inferential Statistics: Lead to inferences (probability that the results would hold true outside of the sample) about the association or relationship between variables in the population.

Associational Research Questions: An associate or relationship between two or more variables usually with each variable having five or more levels.

Assumptions (Statistical): Explains when it is and is not reasonable to perform a specific statistical test.

Attribute (Independent Variable): An independent variable that cannot be manipulated yet is a major focus of the study. A social science research usually a part of who you are (e.g., age, political affiliation, religious believe, etc.).

Bartlett Test: An assumption test for exploratory factor analysis.

Basic (or Bivariate) Statistics: There is *one* independent and *one* dependent variable.

Bivariate Regression: Is used when one wants to predict scores on a normal/scale dependent (outcome) variable from one normal or scale independent (predictor) variable.

Box and Whiskers Plot: A way of displaying the middle 50% in a box and the remaining 50% in whiskers.

Callout Boxes: On the output itself, we have pointed out some of the key things by circling them and making some comments in boxes.

Categorical: Same as a nominal or qualitative variable.

Chi Square: Is a nonparametric test for difference between nominal IV and DV variables. For example, difference in the count between registered Republicans and Democrats across different ethnic group identity.

Codebook or Dictionary: A listing of the characteristics of each variable in one's dataset, including the name of the variable; a longer, more descriptive label for the variable; the meaning of the scores, levels, or values given to data for the variable; etc.

Coding: Is the process of systematically and consistently assigning particular numbers to particular values or levels of a variable.

Cohen's Kappa: A method to assess evidence for interrater *reliability* of a nominal variable.

Complex: Questions and statistics that involve more than two variables at a time; also called multivariate in other texts.

Complex Statistics: There are three or more variables. We decided to call them **complex** rather than **multivariate**, which is more common in the literature, because there is not unanimity about the definition of multivariate, and some statistics with more than two variables (e.g., factorial ANOVA) are not usually classified as multivariate.

Confidence Intervals: If we constructed an infinite number of studies using the same conditions, and computed a 95% confidence interval for each study, 95% of the intervals would contain the true population difference between means. This does not mean there is a .95 probability that the true population difference between means is within our interval. For example, the researcher predicted graduation GPA based upon ACT score and is 95% confident that the student's senior GPA will be between 2.4–3.5.

Content Evidence or Validity: Refers to whether the content that makes up the instrument is a reasonable representation of the concept that one is attempting to measure. Poses the question, does the measure really measure what it is intended to measure?

Correlation Coefficients: These statistics can vary from –1.00 (a perfect negative correlation or association) through .00 (no correlation) to +1.00 (a perfect positive correlation). Note that +1 and –1 are equally high or strong, but they lead to different interpretations.

Correlation Matrix: indicating the associations among all the pairs of three or more variables.

Cramer's V: Measures the strength of a relationship or effect size of two nominal variables when one or both have three or more levels/values.

Critical Value: Found in a statistics table or stored in the computer's memory, that takes into account the degrees of freedom, which are usually based on the number of participants.

Cronbach's Alpha: Is used to provide evidence for the internal consistency reliability of composites, multi-item subscales, or sets of variables resulting from a factor analysis or from a theoretical combination of variables (e.g., several items on a questionnaire designed to measure the same concept).

Dependent Variables: Is assumed to measure or assess the effect of the independent variable. It is the presumed outcome or criterion.

Descriptive Research Questions: These questions are not answered with inferential statistics. They merely describe or summarize data for the sample actually studied, without trying to generalize to a larger population of individuals.

Dichotomous Measure or Variable: A variable with two levels, often dummy coded within SPSS as 0 and 1.

Difference Inferential Statistics: Lead to inferences about the differences (usually mean differences) between groups in the populations from which the samples were drawn.

Difference Research Questions: For these questions, we compare two or more different groups, each of which is composed of individuals with one of the values or levels of the independent variable. This type of question attempts to demonstrate that the groups are not the same on the dependent variable.

Dummy Coding: A special way of coding dichotomous variables, in which 0 = "does not have this characteristic" and 1 = "does have this characteristic". For example, 1 = person in the experimental group, 0 = person in the control group.

Effect Size: The strength of the relationship between the independent variable and the dependent variable, and/or the magnitude of the difference between levels of the independent variable with respect to the dependent variable. An alternative calculation than the significance or p value, which is used to determine how large the impact of a differences or associational finding is. Various statistics calculate effect sizes in different ways. There are many types of effect sizes (e.g., d, r, eta, phi, Cramer's V).

Eta Squared: An effect size given in ANOVA SPSS printouts.

Exploratory Factor Analysis: A common method to test if the constructs a researcher is measuring appear to be measuring the same thing.

Extraneous Variables: These are variables (also called nuisance variables or, in some designs, covariates) that are not of interest in a particular study but could influence the dependent variable.

Environmental factors (e.g., temperature or distractions), time of day, and characteristics of the experimenter, teacher, or therapist are some possible extraneous variables that need to be controlled.

Factor: Another name for group difference, independent variables, or predictor variable.

Factorial ANOVA: One DV and more than one IV, with each IV having two or more levels.

Factorial Evidence: One type of evidence to support the measurement validity of a construct.

Fisher's Exact Test: Should be reported instead of chi-square for small samples if each of the two variables being related has only two levels (2 × 2 cross-tabulation).

Frequency Distribution: A tally or count of the number of times each score on a single variable occurs.

Games-Howell: Used for variables that have unequal variances (Levene's test is significant).

General Linear Model: The relationship between the independent and dependent variables can be expressed by an equation with weights for each of the independent/predictor variables plus an error term.

Homogeneity of Variances: An assumption that the samples in the study have equal variation among the sample.

Independent Samples t-Test: Commonly referred to as just t-test, this is one IV with one normally distributed DV; compares two independent samples for significant differences.

Independent Variables: The variable of cause, difference, or predictor variable.

Inferential Statistics: Statistical procedures used to infer from a sample to a population.

Interquartile Range: The distance between the top and bottom of a box plot, a useful measure of variability for ordinal data or the middle 50%.

Interrater Reliability: A measure of the consistency between two or more raters. For example, the consistency between raters of gymnastic in the Olympics.

Interval: Ordered levels, in which the difference between levels is equal, but no true zero.

Kaiser-Meyer-Olkin Measure: An assumption test run prior to running an exploratory factor analysis.

Kendall's Tau: Measures the strength of the association if both variables are ordinal.

Kruskal-Wallis Test: A nonparametric test similar to one-way ANOVA for ordinal or skewed data.

Kurtosis: The steepness (Leptokurtic) or flatness (Platykurtic) of a curve.

Level of Measurement: The amount of information that numbers represent for that variable. Nominal is the lowest level of measurement, in which numbers just classify people into groups or categories that are not ordered from high to low. Ordinal measures have scores/values that are numbers that can be ordered from low to high, but the amount of difference between adjacent numbers are not clearly equal or are skewed. Interval/Ratio level measures have scores that not only can be ordered from low to high but are more precise, so that differences between adjacent scores should all be equal.

Levene's for Equity of Variance: A statistical test which determines if the variance for two or more groups is approximately the same.

Likert Scale: A response format in which respondents rate themselves or another person using numbers ranging from low to high that correspond to the amount or extent of the characteristic. Often strongly agree to strongly disagree with five or seven levels.

Log Linear Analysis: Is a nonparametric statistic somewhat similar to the between-groups factorial ANOVA for the case where all the variables are nominal or dichotomous.

Logistic Regression: Used for associational questions when the DV is dichotomous. For example, can ACT and high school GPA predict graduation from college or not?

Mann-Whitney: A nonparametric test, which is similar to the independent *t*-test.

MANOVA: (Multivariate analysis of variance) is used if you have two or more normal (scale) dependent variables treated simultaneously.

Maximum: Highest score in a dataset.

Mean: The arithmetic average takes into account all of the available information in computing the central tendency of a frequency distribution.

Measurement: Is the assignment of numbers or symbols to the different characteristics (values) of variables according to rules.

Median: The middle score or median is the appropriate measure of central tendency for ordinal level raw data. The median is a better measure of central tendency than the mean when the frequency distribution is skewed.

Minimum: Lowest score in a dataset.

Missing Value: Variable which is absence from the dataset. A blank value where there should be a record of the datum.

Mixed Design: If the design has a between-groups variable and a within-subjects independent variable.

Mode: The most common category can be used with any kind of data but generally provides the least precise information about central tendency.

Multicollinearity: This effect occurs when there are high inter-correlations among some set of the predictor variables. In other words, multicollinearity happens when two or more predictors are measuring similar information.

Multiple Regression: Which is used to predict a scale/normal dependent variable from two or more independent variables.

Mutually Exclusive: Only one value/score/number can be assigned to each person for the variable.

Negative Correlation: An inverse linear pattern, that is, as X increases, Y decreases.

Negatively Skewed: When extreme scores or the tail of the curve is on the low end; for example, most did well (80–99 on a 100 point exam) but two scores skewed the data distribution with scores of 35 and 28.

Nominal: A level of variable which is categorical, qualitative, or mutually exclusive categories.

Nonexperimental Studies: Uses an attribute independent variable.

Nonparametric Statistics: Statistics designed for ordinal or nominal data and which do not assume normal distribution of data. These tests (e.g., chi-square, Mann-Whitney U, Spearman rho) have fewer assumptions and often can be used when the assumptions of a parametric test are violated. For example, they do not require normal distributions of variables or homogeneity of variances.

Odds Ratio: The odds of one item occurring given the presence of another item. For example, odds of passing a test given previous midterm pass/fail data.

One-Sample t-Test: Compares one group or sample to a hypothesized population mean.

One-Way ANOVA: Compares the *means* of the samples or groups in order to make inferences about the population means. Usually used with three or more group comparisons.

Ordinal: Level of a variable that is ranked and often skewed.

Outliers: Scores that are considered unusually high or low.

Paired Samples Correlations: Provides correlation data between the two paired scores in a paired t-test. That is, used in a repeated measures (within groups) design.

Parametric Statistics/Tests: Statistics designed for normally distributed data.

Pearson Correlation Coefficient, r: A value less than 1.0, varying between −1.0 and +1.0 with 0 representing no effect and +1 or −1 the maximum effect.

Pearson Product Moment Correlation: Measures if the variables are related in a linear (straight line) way. Pearson is measured by r and goes between −1 and +1.

Pilot Study: A small-scale study aimed at testing out procedures and/or measures for a study in order to make needed modifications before engaging in a larger-scale data collection effort.

Positive Correlation: Linear relationship as the X variable increases, so does the Y variable most of the time.

Positively Skewed: When extreme scores or the tail of the curve are on the high end of the distribution and bias the value of the mean.

Post Hoc Test: Tests run after a preliminary tests with three or more groups such as ANOVA and post hoc (e.g., Dunnett's C, Games-Howell, LSD, Scheffe, Tukey & HSD) to determine where the significant finding(s) exist.

Qualitative Variable: In quantitative research, this is the same as a nominal or categorical variable.

Quasi-Experimental: Comparative groups with an active independent variable but not randomly assigned.

Randomized Experimental: Participants randomly assigned to experimental and control group with an active independent variable.

Range: Highest minus lowest score is the crudest measure of variability but does give an indication of the spread in scores if they are ordered and is useful in checking data for errors. For example, a column that is supposed to have only 1s and 0s but the range is 1.3 would indicate an error in the raw dataset.

Ranks: Ordinal variable without a continuous underlying scale.

Ratio: Ordered levels; the difference between levels is equal, and there is a true zero.

Recode: Sometimes it is necessary to recode data, say, from a continuous scale to 1 and 0 indicating a high score or low score or changing raw scores to pass/fail.

Related Samples Design: These designs also include examples where the participants are matched by the experimenter or are related in some other way (e.g., twins, husband and wife, or mother and child).

Repeated Measures Design: When each participant is assessed on the same measure more than once, see within-subjects designs.

Research Hypotheses: Are predictive statements about the relationship between variables, compared to a null hypothesis (sometimes called testable hypothesis), which indicates no difference.

Research Problem: Is a statement that asks about the relationships between two or more variables; however, almost all research studies have *more* than two variables and more than one research question.

Research questions are similar to hypotheses, except that they do not entail specific predictions and are phrased in question format.

Risk Potency Measures: Odds ratios and risk ratios are common examples of a third group or family of effect size measures.

Robust: When the assumption of a statistic can be violated quite a lot without damaging the validity of the statistic.

Rotated Factor Matrix: Is used to find the best position for a three-dimensional set of potential solutions to a factor analysis.

Scale: Interval/ratio variable label used in SPSS and in some textbooks.

Scatterplots: Series of dots made up from each x-y point used in correlations.

Single-Factor ANOVA: A one-way ANOVA comparing one independent variable usually on three or more levels.

Single-Factor Designs: If the design has only one independent variable (either a between-groups design or a within-subjects design).

Skewed: If one tail of a frequency distribution is longer than the other, and if the mean and median are different. A skewness of greater than the absolute value of one as a guide to if parametric or non-parametric statistics should be used.

Spearman Rho: A nonparametric equivalent to Pearson's r when one or both variables are skewed or ordinal.

Standard Deviation: is based on the deviation (x) of each score from the mean of all the scores. Those deviation scores are squared and then summed ($\sum x^2$). This sum is divided by $N-1$, and finally, the square root is taken ($SD = \sqrt{\sum x^2 / N - 1}$).

Standardized Beta Coefficients: Relative weight of each factor in a multiple regression.

String Variable: A variable whose "values" or "levels" are written using letters rather than numbers (e.g., control group vs experimental group, rather than 0 and 1).

Summated Scale: The total scale for a construct measure. For example, your total stress level based upon a ten-item set of stress questions.

Tukey HSD: Used if variances can be assumed to be equal. It is a post hoc test for a significant ANOVA.

Valid Percent: Percentages taking or subtracting the missing values.

Value Labels: Description in words of the meaning associated with different numbers for the variable (e.g., 0 = control group).

Values: A term to describe the several options or categories of a variable; also called categories, levels, and groups.

Variable: A characteristic of the participants or situation in a given study that has different values. A variable must vary or have different variables in the study.

Wilcoxon Signed-Ranks: A non-parametric paired samples or within samples t-test for distributions which are ordinal or skewed.

Within-Subjects Designs: These designs are conceptually the opposite of between-groups designs. In within-subjects designs (also called **repeated measures,** or **related samples** designs), usually a test-retest using the same measure on the same person, for example, your blood pressure last month and now. But can be genetically linked (e.g., your height and your same sex parent height).

Zero Correlation: No linear pattern.

Jessica Bochert

For Further Reading

American Educational Research Association. (2006). Standards for reporting on empirical social science research in AERA publications. *Educational Researcher*, *35*(6), 33–40.

American Psychological Association (APA). (2019). *Concise guide to APA style 7th edition (official)* (7th ed.). Washington, DC: Author.

American Psychological Association (APA). (2020). *Publication manual of the American Psychological Association* (7th ed.). Washington, DC: Author.

Arnau, J., Bendayan, R., Blanca, M. J., & Bono, R. (2013). The effect of skewness and kurtosis on the robustness of linear mixed models. *Behavior Research Methods*, *45*(3), 873–879. https://doi.org/10.3758/s13428-012-0306-x

Baraldi, A. N., & Enders, C. K. (2013). Missing data methods. In T. D. Little (Ed.), *The Oxford handbook of quantitative methods* (Vol. 2, Statistical analysis, pp. 35–663). New York: Oxford University Press.

Baron, R. M., & Kenny, D. A. (1986). The moderator-mediator variable distinction in social psychological research: Conceptual, strategic, and statistical considerations. *Journal of Personality and Social Psychology*, *51*, 1173–1182.

Barrett, K. C., Leech, N. L., & Morgan, G. A. (2025). *SPSS for intermediate statistics: Use and interpretation* (6th ed.). New York: Routledge/Taylor & Francis.

Bishara, A. J., & Hittner, J. B. (2012). Testing the significance of a correlation with nonnormal data: Comparison of Pearson, Spearman, transformation, and resampling approaches. *Psychological Methods*, *17*, 399–417.

Cohen, J. (1988). *Statistical power and analysis for the behavioral sciences* (2nd ed.). Hillsdale, NJ: Lawrence Erlbaum.

Cohen, J. (1994). The world is round (p < .05). *American Psychologist*, *49*, 997–1003.

Enders, C. K. (2013). Dealing with missing data in developmental research. *Child Development Perspectives*, *7*(1), 27–31.

Gliner, J. A., Morgan, G. A., & Leech, N. A. (2017). *Research methods in applied settings: An integrated approach to design and analysis* (3rd ed.). New York: Routledge/Taylor & Francis.

Hair, J. F., Black, B., Babin, B., Anderson, R. E., & Tatham, R. L. (2010). *Multivariate data analysis* (7th ed.). Englewood Cliffs, NJ: Prentice Hall.

Hayes, A. F. (2013). *Introduction to mediation, moderation, and conditional process analysis: A regression-based approach*. Guilford Press.

Ho, A. D., & Yu, C. C. (2015). Descriptive statistics for modern test score distributions: Skewness, kurtosis, discreteness, and ceiling effects. *Educational and Psychological Measurement*, *75*(3), 365–388. https://doi.org/10.1177/0013164414548576

Hox, J. J. (2013). Multilevel regression and multilevel structural equation modeling. In T. D. Little (Ed.), *The Oxford handbook of quantitative methods* (Vol. 2, Statistical analysis, pp. 81–293). New York: Oxford University Press.

Huck, S. J. (2011). *Reading statistics and research* (6th ed.). Upper Saddle River, NJ: Pearson.

Kelley, K. (2005). The effects of nonnormal distributions on confidence intervals around the standardized mean difference: Bootstrap and parametric confidence intervals. *Educational and Psychological Measurement, 65*(1), 51–69. https://doi.org/10.1177/0013164404264850

Little, R. J. A., & Rubin, D. B. (2002). *Statistical analysis with missing data* (2nd ed.). Hoboken, NJ: Wiley.

Morgan, G. A., Gliner, J. A., & Harmon, R. J. (2006). *Understanding and evaluating research in applied and clinical setting.* Mahwah, NJ: Lawrence Erlbaum.

Morgan, S. E., Reichart, T., & Harrison, T. R. (2016). *From numbers to words: Reporting statistical results for the social sciences.* Boston, MA: Allyn & Bacon.

Newton, R. R., & Rudestam, K. E. (2013). *Your statistical consultant: Answers to your data analysis questions* (2nd ed.). Thousand Oaks, CA: Sage.

Nicol, A. A. M., & Pexman, P. M. (2010a). *Displaying your findings: A practical guide for creating figures, posters, and presentations.* Washington, DC: American Psychological Association.

Nicol, A. A. M., & Pexman, P. M. (2010b). *Presenting your findings: A practical guide for creating tables.* Washington, DC: American Psychological Association.

Raudenbush, S. W., & Bryk, A. S. (2002). *Hierarchical linear models: Applications and data analysis methods* (2nd ed.). Thousand Oaks, CA: Sage.

Rudestam, K. E., & Newton, R. R. (2007). *Surviving your dissertation: A comprehensive guide to content and process* (3rd ed.). Newbury Park, CA: Sage.

Singer, J. D. (1998). Using SAS PROC MIXED to fit multilevel models, hierarchical models, and individual growth models. *Journal of Educational and Behavioral Statistics, 23,* 323–355.

Snijders, T. A. B. (2005). Power and sample size in multilevel modeling. In B. S. Everitt & D. C. Howell (Eds.), *Encyclopedia of statistics in behavioral science* (Vol. 3, pp. 570–1573). Chichester: Wiley.

Tabachnick, B. G., & Fidell, L. S. (2007). *Using multivariate statistics* (5th ed.). Boston, MA: Allyn & Bacon.

Thompson, B. (Ed.). (2003). *Score reliability: Contemporary thinking on reliability issues.* Thousand Oaks, CA: Sage.

Thompson, B. (2004). *Exploratory and confirmatory factor analyses: Understanding concepts and applications.* Washington, DC: American Psychological Association.

Vaske, J. J., Gliner, J. A., & Morgan, G. A. (2002). Communicating judgments about practical significance: Effect size, confidence intervals and odds ratios. *Human Dimensions of Wildlife, 7,* 287–300.

Wilkinson, L., & The APA Task Force on Statistical Inference. (1999). Statistical methods in psychology journals: Guidelines and explanations. *American Psychologist, 54,* 594–604.

Wuensch, K. L., & Poteat, G. M. (1998). Evaluating the morality of animal research: Effects of ethical ideology, gender, and purpose. *Journal of Social Behavior & Personality, 13,* 139–150.

Index

Note: Commands or terms used by SPSS but less common in statistics or research methods books are in **bold**